Cellular Analysis by Atomic Force Microscopy

Cellular Analysis by Atomic Force Microscopy

Malgorzata Lekka

Pan Stanford Publishing

Published by

Pan Stanford Publishing Pte. Ltd.
Penthouse Level, Suntec Tower 3
8 Temasek Boulevard
Singapore 038988

Email: editorial@panstanford.com
Web: www.panstanford.com

British Library Cataloguing-in-Publication Data
A catalogue record for this book is available from the British Library.

Cellular Analysis by Atomic Force Microscopy

ISBN 978-981-4669-67-2 (Hardcover)
ISBN 978-1-315-36480-3 (eBook)

Printed in Canada

Contents

Preface

In the recent two decades, there has been ample evidence that shows the capability of detecting pathologically changed cells basing on their mechanical properties. Such studies commenced to be attractive for biology and medicine with the development of techniques enabling the measurements at a single-cell level. One of such techniques is atomic force microscopy (AFM). Despite evidence showing the feasibility of AFM to identify cells with altered elastic and adhesive properties, the use of this technique as a complementary diagnostic method, for example, in cancer treatment, raises a lot of controversy. This is primarily due to the still high complexity of the apparatus, the complexity of the experiment and data analysis, and the lack of basis for the rapid verification of the results. In addition, there are also doubts whether a high sensitivity of the method (measuring the properties of a single cell) will be sufficient to validate the cancer cell identification at the same level as it is carried out in the analysis of aspirate composed of millions of cells at different stages of progression.

The monograph *Cellular Analysis by Atomic Force Microscopy* presents the use of the AFM as a tool for the characterization of cancerous cells by studies of cellular deformability and cells' adhesive properties. The text is organized as follows: after a brief introduction pointing out the significance of cellular deformability and adhesive properties, Chapter 2 shortly discusses the basic elements of the cell structure, on the knowledge level that facilitates understanding of the role and significance of mechanical properties in cancer. Chapter 3 is devoted to the AFM technique itself. It covers the description of the idea and the construction details of the atomic force microscope, introduces the force spectroscopy mode of its operation, and discusses the calibration issues. The next two chapters are focused on specific applications: Chapter 4 is devoted to cellular deformability measurements by the AFM and Chapter 5 introduces the technique for quantification of cellular adhesive properties. The monograph concludes with

Final Conclusions, briefly summarizing the use of atomic force microscopy in cancer research.

This monograph* is designed to be a kind of a practical textbook, presenting in a concise manner the methodology of assessing the mechanical characteristics of individual cells by AFM. It is addressed to those researchers who need some hints in cases when strange results came out from the AFM experiment. I hope that the monograph will help eliminate at least some questions.

I would like to dedicate this book to all people who helped me in different ways in my research. I am taking this moment to express my deep gratitude to all of them. In particular, I would like to mention Zbigniew Stachura (IFJ PAN), who was always ready to answer questions and give invaluable advice, and whose continuous support enabled to establish and organize the biological activity of the AFM laboratory; Piotr Laidler (CM UJ) for teaching me how to understand biological phenomena, for his help during the realization of my biological experiments, and, most important, for showing me the importance of my studies; Jan Styczeń (IFJ PAN) for his support and for encouraging me in my studies over many years during his heading of the former Department of Nuclear Spectroscopy. I would like to also acknowledge Joanna Wiltowska-Zuber (IFJ PAN) for her invaluable help with managing the AFM laboratory, especially during the time when I was writing this monograph. Finally, I would like to thank my whole family and especially my husband, Janusz, for continuous support.

The topics of this book were also conceived within the networking research activity of the EU COST Action on Applications of Atomic Force Microscopy to NanoMedicine and Life Sciences (AFM4NanoMed&Bio) and the NCN project number NCN DEC-2011/01/M/ST3/00711.

Malgorzata Lekka

*A large portion of the this book has been prepared on the basis of my IFJ PAN Report, 2001/AP entitled "The use of atomic force microscopy as a technique for the identification of cancerous cells."

Chapter 1

Introduction

Cancer is a very complex disease, involving multiple molecular and cellular processes arising from a gradual accumulation of genetic changes in individual cells. The most apparent morphological change is visible during the transition from a benign tumor to metastatic tumors where cells alter from highly differentiated normal phenotypes to migratory and invasive ones. Around 90% of all cancer deaths are due to metastatic spread of primary tumors. The criteria utilized to detect cancerous cells have been mainly relying on biological and morphological description, additionally complemented by a variety of other techniques, including genetic, chemical, and immunological methods, applied in order to fine-tune diagnosis or therapy. Despite enormous efforts to develop better treatment protocols, our ability to cure solid tumors, such as those of the breast, prostate, cervix or colon, is still lacking sufficient detection methods [1].

The cells transformed oncogenically differ from normal ones in many ways, encompassing variations in any cellular aspects such as growth, differentiation, interactions between neighboring cells and/or with the extracellular matrix (ECM), cytoskeleton organization, and several others. Poor differentiation of the cytoskeleton can result in the larger deformability of cancerous cells. Low stiffness of cancer cells is related to a partial loss of actin filaments and/or microtubules, and therefore by lower density of

Cellular Analysis by Atomic Force Microscopy
Malgorzata Lekka

ISBN 978-981-4669-67-2 (Hardcover), 978-1-315-36480-3 (eBook)
www.panstanford.com

the cellular scaffold [2, 3]. Moreover, one of the key phenomena in metastasis includes adhesive interactions, maintained by distinct type of adhesion molecules present on a cell surface. Cancerous cell aptitude for invasion and migration (clinically interpreted as tumor aggressiveness) has been associated with poor differentiation of the cell and the altered adhesive interactions that characterize a vast majority of cancer cells.

It is obvious that novel techniques are in the limelight if they are able to bring more precise, local information about cancerous changes as early as possible. There are rather few methods capable to assess cell mechanical properties. Historically, the first technique was the micropipette aspiration [4, 5]. Other researchers have employed the magnetic bead rheology [6], microneedle probes [7], acoustic microscopes [8], and the manipulation of beads attached to cells with optical tweezers [9, 10]. Among these techniques, the atomic force microscopy (AFM) can detect malignant changes with a very high resolution, being applied either in imaging mode or as the technique providing information about the mechanical properties of living cells (i.e., their ability to deform and to adhere) in a quantitative manner. Its main advantage is the possibility to measure biological objects directly in their natural environment, such as buffer solutions or culture media.

Many publications in this area were devoted to the characterization of single cells' deformability and adhesiveness, presented in a broad context of biological targets, starting from cell motility, would healing, muscle contraction or differentiation and ending up in characterization of various pathologies such as muscular dystrophies, blood diseases or cancers. Therefore, in this Chapter, the importance of cellular ability to deform and to adhere is presented with the focus on the AFM-related aspects in cancer studies.

1.1 Cell Ability to Deform

Within the past two decades, the cellular ability to deform has attracted great interest in the field of biology. This is because in human body, various cell types are continuously exposed to passive (stretch, compression) and/or active (contraction) deformations. The technological development of techniques, that enable to probe elastic properties of single cells, has been shown to be more powerful than that of bulk measurements since the former ones can relate the mechanical properties with cellular functioning and structures.

The capability of cells to deform has been studied long time ago. One of the earliest reports of increased deformability of cancerous cells has been reported by Ochalek et al. [11]. In these studies, the microfiltration experiments were used to study the migration capability of B16 mouse melanoma cells. In the filtration experiment, the assumption was that all melanoma cells capable to metastase passed through the filter. This was justified by the condition that metastatic cells must be squeezed to go through the surrounding tissue matrix when they make their way into the circulatory systems where they are directed to establish distant settlements. However, the conclusion has been built by counting cells and quantifying the filtration time, not by the determination of cells mechanical properties. The pioneering study [12] showed the importance of mechanical properties to characterize cancerous cells. In these studies, the deformability of human bladder cancerous cells (cell lines: T24, Hu456, BC3726) was one order of magnitude larger than for their reference counterpartners (cell lines: Hu609, HCV29). These early results have been supported (and indirectly verified) by optical tweezers measurements. Using this latter, high throughput technique, three cell lines were compared, namely, a non-tumorigenic breast epithelial MCF10 cells, a non-motile, non-metastatic breast epithelial cancer MCF7 cells, and MCF7 cells transformed with phorbol ester causing the increase in the cancer cell invasiveness. The results showed significant increase of MCF7 cells deformability compared to MCF10 and non-transformed MCF7 ones [10].

Based on single-cell deformability measurements, it has been found that cell structure is closely related to specific mechanical properties, which, in turn, depend on the organization of cell cytoskeleton. The role of cytoskeletal components (mainly actin filaments and microtubules) in cellular deformability has been shown by applying so-called cytoskeletal drugs that influence the structure and formation of each particular component [13–15]. For example, cytochalasin D increases the cellular deformability while nocodazol leads to cell stiffening. A summary of cytoskeletal drugs influence on cellular deformability is presented in

Table 1.1. Depending on the type of the compounds, disrupting or stabilizing particular cytoskeletal elements, the influence on cellular deformability manifests either in higher or in lower deformability (cells become softer or more rigid, respectively). However, it should be underlined that the effect is dependent on the applied concentration and time of action.

Table 1.1 The effect of common cytoskeletal drugs on cellular deformability

Drug	Known effect	Effect on cellular mechanics
Cytochalasin B	Disrupt actin filaments (F-actin) Disassembled stress fibers	Increased deformability
Cytochalasin D	Disrupt actin filaments (F-actin) Disassembled stress fibers Aggregation of actin within the cytosol	Increased deformability
Latrunculin A	Disrupt actin filaments (F-actin) Disassembled stress fibers	Increased deformability
Jasplakinoide	Disrupt actin filaments (F-actin) but did not disassemble stress fibers	Increased deformability (cell becomes softer)
Colchicine	Disrupt microtubules	No effect
Colcemid	Disrupt microtubules	No effect or increased deformability
Taxol (paclitaxel)	Stabilize microtubules	No effect or decreased deformability
Nocodazol	Stabilize microtubules	No effect or decreased deformability

The cytoskeleton interaction with associated proteins has been demonstrated to influence cellular elastic properties for cells expressing vinculin (a focal adhesion protein interacting with actin fibers). The loss of vinculin reflects in a noticeable reduction of cell adhesion, spreading and the presence of stress fibers. The comparison performed by Goldman et al. in 1998 showed that the vinculin-deficient F9 mouse embryonic carcinoma cells had lower Young's modulus than the wild-type cells. The authors attributed these changes to altered actin cytoskeletal organization, indicating an important role of vinculin as an integral part of the cytoskeletal network [16].

In the AFM technique, the deformability is expressed by Young's modulus value which delivers a quantitative measure of cellular elasticity. It is a very local feature, showing usually large discrepancy when measured on a single cell as well as within a population of cells. The latter variations are attributed to heterogeneity of cellular structures, while the former reveal a non-uniformity of cell populations. It has been reported that cells *in vitro* have Young's modulus values in the range of 1–100 kPa [17, 18], which encompasses different types of investigated cells, including vascular smooth muscle cells, fibroblasts, bladder cells, red blood cells, platelets, and epithelial cells. Since different cell types are measured, the large modulus variation is fully justified. On the other hand, one should remember that the determination of Young's modulus by the AFM technique delivers only its relative value; therefore, it can be used only for comparative studies where all experimental conditions are conserved (see Chapter 3 for details). However, the gathered evidences enable to make a general statement that malignant/metastatic cells respond either more elastically or less viscously to the applied stress. These findings suggest that cell stiffness can be used as a marker, and, also, as a diagnostic parameter for the underlying disease (Fig. 1.1).

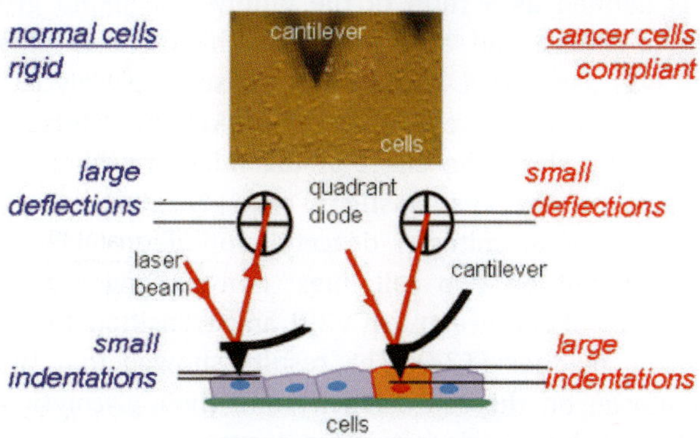

Figure 1.1 The idea of cancer cell recognition based on the AFM elasticity measurements. When normal cells are measured, the resulting indentation depths are smaller due to cells' rigidity. On the contrary, cancerous cells manifest in larger indentation depths, indicating their larger deformability compared to normal cells. Reprinted with permission from [19].

Despite the lack of the absolute value of Young's modulus, the obtained relative change of the elastic modulus is sufficient to describe the alterations in cancerous cells and to monitor its changes due to various effects. The cell stiffness of cancerous cells appears to be non-specific and induced within a whole volume of a cell. Therefore, it seems to be easier measurable by different techniques, including the AFM. The measurements of Young's modulus, carried out on single cellular level, can help to determine the range of cytoskeleton changes and allow their quantification. The following subsections present the examples of the use of cancer cell deformability to describe the influence of drugs, the sensing of substrate stiffness, and the correlation of cellular deformability with the malignancy degree.

1.1.1 Monitoring Chitosan Effect on Cancerous Cells

The capability to measure the cellular deformability enables to monitor changes occurring within cell populations under the influence of chemical compounds added to culture medium. This has been demonstrated in studies showing a correlation between cell stiffness and glycolytic activity as a function of various chitosan deacetylation degrees [20]. The deacetylation degree is defined as a ratio of the number of amino groups to the sum of amino and acetyl groups. The increase of Young's modulus was associated with the decrease of glycolytic activity in cancerous cells measured by lactate and ATP levels (Fig. 1.2).

The results show the effect of the microcrystalline chitosan preparations on the cell elasticity and glycolytic activity as a function of various chitosan deacetylation degree. The studies were carried out on two cell lines: non-malignant transitional epithelial cells of the ureter (HCV29) and transitional cell cancer of the urine bladder (T24). The results showed that the effect of the chitosan on the cell elasticity and their glycolytic activity is a function of chitosan deacetylation degree.

The metabolism of cancerous cells shows the high activity and it is usually connected with the overexpression of the glycolytic enzymes like, pyruvate kinase type M2 [21, 22]. Some of glycolytic enzymes are present either in cytosol or associated with the cytoskeleton. The detachment of the cytoskeleton-associated enzymes from the cytoskeleton leads to the decrease in the level of the glycolysis and to reorganization of cell cytoskeleton by the rearrangement in the actin or microtubules network [22].

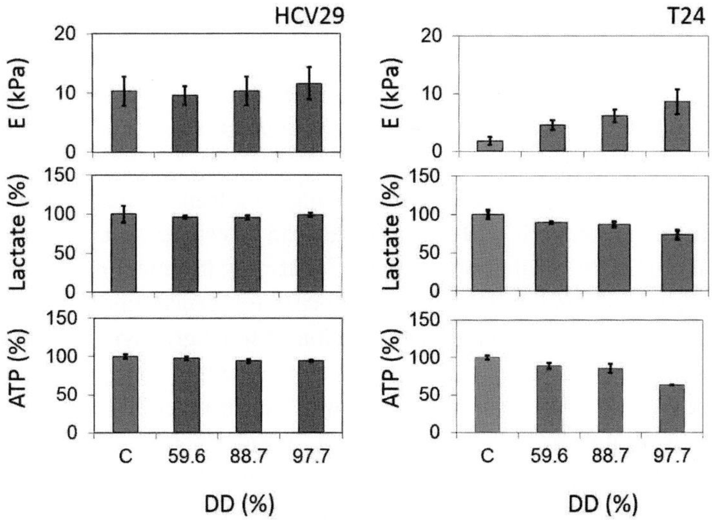

Figure 1.2 The effect of chitosans on elastic properties and glycolytic activity of non-malignant HCV29 cells (left panel) and cancerous T24 ones (right panel). C denotes the control, i.e., non-treated cells while DD is the deacetylation degree expressed in %. The glycolytic activity was measured as a lactate and ATP level normalized to control. Adapted from [20].

The deacetylation degree can be attributed to the charge proportional to the amount of amino groups, which are positively charged in water solutions. The high molecular mass of chitosan does not allow it to enter the cell and therefore its interaction with cells is restricted to cell membrane only. This interaction is probably caused by binding the positively charged molecules of chitosan—the higher is its deacetylation degree, the higher positive charge—to the negatively charged cell membrane. Such a mechanism is suggested by the stronger interaction between the chitosan with the highest deacetylation degree and cell surface. Significant and large changes were observed for cancerous cells in all three types of measurements: lactate production, ATP level, and Young's modulus values. Reference cells behave differently. The level of lactate and ATP was almost similar, regardless on the deacetylation degree (i.e., charge); only a weak relation was visible in Young's modulus values, indicating some chitosan interaction with membrane of these cells. This difference between HCV29 and T24 cells behavior can be explained by assumption of the smaller negative charge on the surface of the non-malignant cells compared to the latter ones. Therefore, the reference cells are less covered with positively charged chitosan and chitosan influence is lower.

The effect observed for non-malignant transitional epithelial cell of ureter (HCV29) and transitional cell cancer of urine bladder (T24) cell lines suggests that the interaction between the chitosan and cell surface is specific for cancerous cells. This can be attributed to the overexpression of some glycolytic enzymes, specific for cancerous state, like for example the isoenzyme pyruvate kinase type M2. Chitosan can interact in two ways. Either chitosan interacts with the actin cytoskeleton and by changes in its structure it can inhibit the process of glycolysis or chitosan interacts with glycolytic enzymes and by inhibition of glycolysis it can cause the reorganization of actin cytoskeletal network.

1.1.2 Mechanosensitivity of Cancerous Cells

Over past years, a great effort has been made to understand the influence of substrate stiffness on behavior of living cells. The elastic properties of various tissues in living organisms vary from few Pa for very soft tissues like brain, to tens of kPa in muscles, and even to MPa for some cartilages [23]. To study the effect of the mechanical properties on single cells, hydrogels, such as polyacrylamide or collagen ones, are used to mimic cellular environment with stiffness within the range from 10 Pa to hundreds of kPa. The effect of substrate stiffness on living cells properties, growth, and differentiation has been demonstrated mostly for normal cells. However, recently, several studies have shown the influence of substrate stiffness on cancerous cells properties (Table 1.2).

To study the effect of the environment, cells are usually cultured on substrates coated with distinct ECM proteins (mostly fibronectin, collagen, and laminin).

Table 1.2 Some examples of mechanosensitivity of cancerous cells (PAA denotes the polyacrylamide gels)

Cell type	Substrate	Ref.
Prostate cancer PC-3	2D collagen matrix Collagen coated PAA	[24]
Human colon cancer (HCT-8)	Fibronectin coated PAA	[25]
Human colon cancer (HCT-8)	Fibronectin coated PAA	[25]
non-cancerous, human-derived MCF10A MEC cells	Collagen matrix	[26]
Prostate cancer (PC-3 & Du145) Melanoma cells (WM115 & WM266-4)	Poly(dimethyl)siloxane (PDMS)	[27]

Cancer cells have been shown to possess altered mechanosensitivity manifested in distinct mechanical response to various substrate stiffness. The altered mechanosensitivity of cancerous cells is attributed to their mesenchymal phenotype, which they acquire during cancer progression. One of the few examples showing such mechanical response to substrate stiffness is presented below (Fig. 1.3).

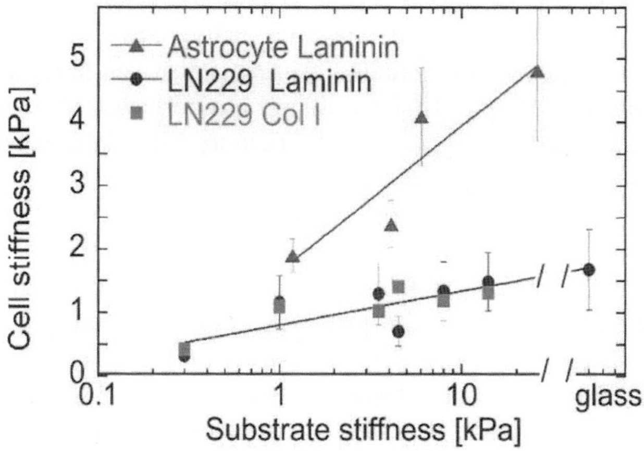

Figure 1.3 Glioma cells LN229 response to substrate stiffness: polyacrylamide gels covered either with laminin or collagen (Col I) compared to normal astrocytes cultured on laminin. Reprinted from [28].

For cells cultured on a glass substrate, glioma cells are softer than normal astrocytes. Their elastic modulus changes as a function of polyacrylamide stiffness but the character of the response is different compared to normal astrocytes. For glioma cells, the elastic modulus changes from around 0.5 to 1.5 kPa, while, in contrast, normal astrocytes are stiffer on all gel substrates reaching the elastic moduli over 4 kPa compared to initial 2 kPa for cells grown on the softest gel substrates. Interestingly, the presence of two distinct ECM proteins (collagen and fibronectin) does not change the mechanical response of these cells [28]. The response of glioma cells to substrate stiffness suggests that the observed cellular stiffness increase might contribute to augmented tension, motility, or proliferation, by analogy with other malignancies such as breast cancer [29].

1.1.3 Stiffness as Cancer Grades

Having in mind the potential use of cellular elasticity as a biomarker of cancer-related changes, the finding of the correlation between cancer grades and elastic properties is of a great interest for oncology. So far, few papers reported that such relation exists for ovarian cancer [30]. Analogously to other cancer cells, the non-malignant IOSE cells show larger cellular stiffness than two other studied ovarian cancer cell lines, namely, HEY and HEY A8 ones (these cell lines are derived from the same tumor specimen). The HEY A8 cells are more deformable compared to HEY cells (Fig. 1.4).

On the other hand, the studied cells also differ in their tumorigenicity in nude mice whereby HEY A8 cells are more tumorigenic after intraperitoneal injection to nude mice [31]. The migratory and invasive properties of both studied HEY and HEY A8 cell display the largest invasive and migratory activity for HEY A8 cells and the lowest for IOSE control cells, indicating that cellular stiffness is inversely correlated with the indicators of metastatic potential (migration and invasiveness).

During cancer progression, ovarian cancer cells acquire the ability to penetrate into blood vessels and to migrate to distant tissues where they form another tumor site. In order to relocate, metastatic cancer cells must undergo deformations, which is possible due to irregular and disorganized actin microfilament network (determining cell's mechanical properties). Therefore, quantification of cellular stiffness in ovarian cancers could be used to distinguish them from benign cells. However, further studies are required, using cancer cells obtained from patients to fully gauge the usefulness of elastic properties as a biomarker.

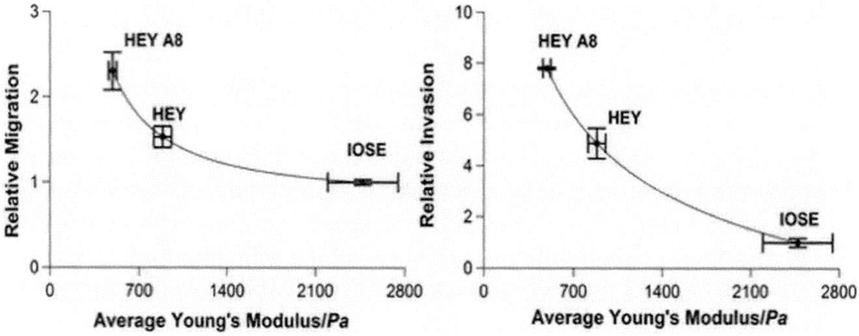

Figure 1.4 Relative migration and invasion plotted as a function of average stiffness for ovarian IOSE, HEY, and HEY A8 cancer cells. For clarity, data points are fitted with power law. Error bars: standard errors of the mean values. Reprinted from [30].

1.2 Cell Ability to Adhere

Cell adhesion plays a crucial role in many events governing the maintenance of various tissue structures and integrity, where the interactions between cells themselves and also between cells and their environment are formed. Although these adhesive interactions seem to be stable, they should be considered as dynamic ones, since in many cases the binding/unbinding events are intimately involved both in changes of cell arrangements responsible for various tissue structures and in proper physiological functioning of cells [32]. The defensive role of the immune system is an example of the process where the interactions between single molecules are essential for organisms. Malfunctioning of the immune system can result in severe disorders, such as autoimmune diseases, hypersensitivities, or immune deficiency. Another example of the cell adhesion is a cell migration, which is dependent on the continuous formation and dissociation of specific bonds between the adhesion molecules and the environment [33, 34]. All these biological functions of proteins depend on their direct physical interactions (i.e., contacts) with other molecules.

The cancerous transformation changes adhesive properties, therefore, a natural direction for the studies is a determination of changes in cell's adhesive interactions through measurements of unbinding forces between particular molecules that are well-established (as potential) markers of the disease. The expression of different types of molecules can be studied using several other methods. The most popular are the immunodetection of blotted proteins (i.e., Western blot technique) and fluorescence based methods (fluorescence microscopy or flow cytometry). In the former method, the presence of a given interaction can be detected and used for estimation of the molecular mass of the formed complexes.

The atomic force microscopy (AFM) working in force spectroscopy mode has become an essential tool to study single molecule and /or single-cell interactions in a quantitative manner [35]. The AFM measurements of the unbinding forces are of great interest for many biologists, since in contrast to standard biochemical or cell biology methods, this technique can be used not only for the detection of molecules on the cell surface but also it gives additional information about the physical properties of the studied interaction. The technique offers a convenient way to measure forces involved in specific interactions by detection of unbinding events between the AFM tip, functionalized with one type of molecules (a ligand), probing its complementary partner (a receptor) which can be either immobilized on the substrate or present on a surface of the cell plasma membrane. So far, the AFM technique has been mostly applied to study the interaction of different molecular complexes for isolated molecules in buffer conditions [36].

1.2.1 Specific Interactions in Living Cells

Some attempts have been also made to investigate specific interactions directly on a surface of living cells. There are two approaches to realize this—either the probing AFM cantilever is functionalized with a ligand specifically recognizing the corresponding surface receptor or a single cell is used as a probe (see Chapter 5 for details). The latter approach requires that the force needed to uproot a single receptor from the cell membrane is much smaller than the force necessary to break the molecular bonds [37]. This assumption has been justified by several studies [38, 39].

Depending on the way of the data collection, one can distinguish two modes of measurement: static and dynamic one. The static force spectroscopy, when performed at well-defined experimental conditions, can provide useful information about the strength of the interaction of different types of molecules present on a surface of living cells. In that manner the interaction between integrins present on living osteoclasts and their different ligands was quantified [39]. The studied ligands had a common RGD (Arg-Gly-Asp) sequence which was recognized by integrins with high affinity. The obtained unbinding force varied from 32 to 97 pN, demonstrating that the context of RGD sequence within a protein considerable influences the unbinding force value.

Several studies have been focused on differences between normal and cancerous cells where the AFM technique has been shown to be complementary to other biochemical methods [40–42]. Using AFM, the characterization of the expression of cell surface molecules in different cancerous cells can be applied to show the antigenic identity of receptors or their dissimilarities or alterations of the binding site within the same receptor type.

During the progression of prostate cancer the expression of prostate-specific antigen (PSA) increases. Unfortunately, during inflammation process, the level of PSA also raises leading to misdiagnosis of prostate cancers. To improve this, new markers are searched for early and accurate staging as well as better means to treat and follow the progress of therapy. One such a potential marker is prostate-specific membrane antigen (PSMA). Surprisingly, some androgen-independent metastatic prostate cancer cell lines, like PC-3 and Du-145, do not express this antigen on the surface. The re-expression of PSMA can be restored by cells under incubation with a specific growth factor, which can be detected using blotted proteins approach. The drawback is to detect the overall prostate-specific antigen independently of its form (membranous or cytosolic one). The use of the AFM limits the measurements only to fraction of antigens present on cell surface, i.e., only to the membrane form of prostate-specific antigen [40]. Moreover, it has been shown that the induction of PSMA expression by the chosen growth factors, as estradiol (E) and basic fibroblast growth factor (bFGF) does not change significantly the unbinding force, revealing binding site similarity before and after restoration of PSMA expression (Table 1.3).

Table 1.3 The unbinding force measured for the interaction between monoclonal antibody against PSMA present on a surface of living prostate cell

Cell line	Conditions	Unbinding force
LNCaP	Control	63 ± 27 pN
	bFGF stimulated	44 ± 17 pN
	E stimulated	58 ±25 pN
PC-3	Control	61 ± 23 pN
	bFGF stimulated	45 ± 17 pN
	E stimulated	65 ± 16 pN
Du145	Control	65 ± 27 pN
	bFGF stimulated	50 ± 22 pN
	E stimulated	60 ± 14 pN

These studies showed systematically smaller values of the unbinding forces upon PSMA stimulation with bFGF, independently of the type of cancer cell lines.

The differences in single-molecule interactions in living cancerous cells have been demonstrated for carbohydrate interactions studied directly on human bladder cell lines [41]. This work proposed the AFM as a method to interrogate both the identity of the complex carbohydrates, as well as their relative number on the cell surface. The unbinding force, determined for the same molecular ligand, suggested slightly dissimilar structure of the binding sites of the same receptor present on cell surface. For the other lectin probe (agglutinin from *Phaseolus vulgaris*), much larger unbinding force indicated a distinct structure of the binding site in cancerous cells. The unbinding probability confirmed a higher content of both sialic acids and mannose-containing carbohydrates in cancerous and reference cells.

Another example are studies on alterations in the expression of cell–cell adhesion molecules, i.e., N-cadherin, whose expression increases in more advanced stages of cancer progression (Fig. 1.5).

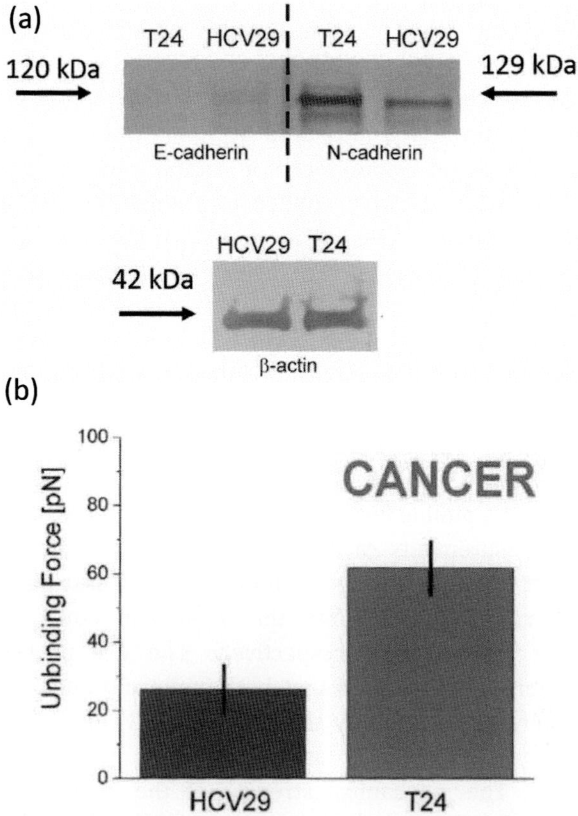

Figure 1.5 The expression of N-cadherin in human bladder cancer cells (HCV29—non-malignant cell cancer of ureter and T24—transitional cell carcinoma). (a) Western blot showing the increase of N-cadherin followed by decrease of E-cadherin normalized to β-actin expression level. (b) The unbinding force determined using AFM. Adapted with permission from [42].

The unbinding force between N-cadherin and the monoclonal antibody (GC4) is expected to have the same value if the structure of the N-cadherin binding site remains unchanged upon cancer transformation. However, the obtained experimental results clearly suggest otherwise. The unbinding force studied for N-cadherin present in cancerous cells (T24) is roughly 2.5 times larger compared to N-cadherins probed on a surface of reference HCV29 cells, namely 61.7 ± 14.6 pN versus 26.1 ± 7.1 pN, respectively (the comparison is possible under constant loading rate; see Chapter 5 for details). These results suggest the altered structure of the cadherin binding site.

The unbinding force measured by AFM is not a fundamental property of a single ligand–receptor complex but it depends on how fast the external force applied to a single bond changes in time (i.e., on loading rates). The studies of the unbinding force as a function of loading rate bring the information on the properties of the energy landscape of the interacting molecules. Such studies, called dynamic force spectroscopy, were applied to investigate the single-molecule interactions on a living cell surface. The most common integrin receptor is the $\alpha 5\beta 1$ integrin, which interacts with a fibronectin (FN). An important intrinsic property of the $\alpha 5\beta 1$–FN interaction is the dynamic response of the complex to a pulling force (Fig. 1.6).

AFM measurements carried out for this interaction revealed two distinct regimes during unbinding: both a fast-loading rate and a slow one characterizes the inner and outer activation barriers of the complex, respectively. The use of genetically modified fibronectin showed that both inner and outer activation barriers were suppressed by the mutation [43]. In other studies, the adhesion mediated by another integrin ($\alpha 4\beta 1$) has been investigated. The mechanical strength of the interaction allows resisting to the large shear forces imposed by the bloodstream [44]. The employed single-molecule dynamic force spectroscopy showed that the dissociation of the $\alpha 4\beta 1$ from its ligand complex involves overcoming of at least two energy barriers: a steep inner barrier and a more elevated outer one. The inner barrier grants the complex the tensile strength to withstand large pulling forces (>50 pN). The outer barrier of the complex was stabilized by integrin activation. Together, these findings can provide a molecular explanation for the functionally relevant kinetic properties of the studied interaction involving $\alpha 4\beta 1$ integrin.

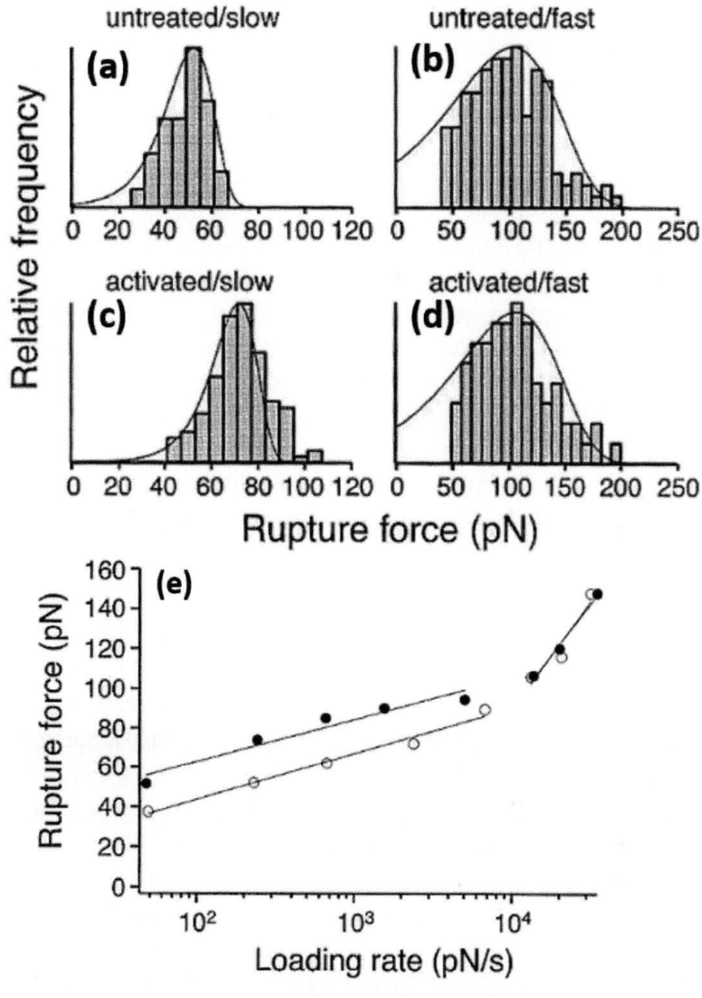

Figure 1.6 Measurements of the rupture force of individual $\alpha 5\beta 1$ integrin–fibronectin complex. The obtained force histograms show unbinding events distributions recorded at low (200–250 pN/s) and high (13000–13500 pN/s) loading rates for untreated (a, b) and activated by 10% TS2/16 (c, d) K562 cells. The fitted probability density functions were obtained using the Bell model parameters (see Chapter 5 for details). (e) The Bell model (lines) fitted to the rupture force measured for untreated (open circles) and activated (solid circles) K562 cells as a function of loading rate. Reprinted with permission from [43].

Another studies by, Taubenberger et al. characterized early steps of integrin-mediated cell adhesion to a collagen type I by using single-cell force spectroscopy [45]. In agreement with the role of $\alpha 2\beta 1$ integrin as a collagen type I receptor, the $\alpha 2\beta 1$-expressing Chinese hamster ovary (CHO)-A2 cells spread rapidly on the matrix, whereas $\alpha 2\beta 1$-negative CHO cells adhere poorly. Probing CHO-A2 cell detachment forces over a contact time of several tens of seconds revealed a nonlinear adhesion response. During the first 60 s, cell adhesion increased slowly, and forces associated with the smallest rupture events were consistent with the breakage of individual integrin–collagen bonds. Above 60 s, a fraction of cells rapidly switched into an activated adhesion state, marked by up to 10-fold increased detachment forces. Elevated overall cell adhesion coincided with a rise of the smallest rupture forces above the value required to break a single integrin–collagen bond, suggesting a change from single to cooperative receptor binding. Transition into the activated adhesion mode and the increase of the smallest rupture forces were both blocked by inhibitors of actomyosin contractility.

References

1. Christofori, G. (2006). New signals from the invasive front, *Nature*, **441**, pp. 444–450.
2. Ben–Zéev, A. (1997). Cytoskeletal and adhesion proteins as tumor suppressors, *Curr. Opin. Cell Biol.*, **9**, pp. 99–108.
3. Yamazaki, D., Kurisu, S., and Takenawa, Y. (2007). Regulation of cancer motility through actin reorganization, *Cancer Sci.*, **96**, pp. 379–386.
4. Drury, J. L., and Dembo, M. (1999). Hydrodynamics of micropipette aspiration, *Biophys. J.*, **76**, pp. 110–128.
5. Hochmuth, R. M. (2000). Micropipette aspiration of living cells, *J. Biomech.*, **33**, pp. 15–22.
6. Bausch, A. R., Möller, W., and Sackmann, E. (1999). Measurements of local viscoelasticity and forces in living cells by magnetic tweezers, *Biophys. J.*, **76**, pp. 573–579.
7. Zahalak, G. I., and Ma, S. P. (1999). Muscle activation and contraction: Constitutive relations based directly on cross-bridge kinetics, *J. Biomech. Eng.*, **112**, pp. 52–62.
8. Kundu, T., Lee, J. P., Blasé, C., and Bereiter-Hahn, J. (2006). Acoustic microscope lens modeling and its application in determining

biological cell properties from single- and multi-layered cell models, *J. Acoust. Soc. Am.*, **120**, pp. 1646–1654.

9. Sleep, J., Wilson, D., Simmons, R., and Gratzer, W. (1999). Elasticity of the red cell membrane and its relation to hemolytic disorders: And optical tweezers study, *Biophys. J.*, **77**, pp. 3085–3095.
10. Guck, J., Schinkinger, S., Lincoln, B., Wottawah, F., Ebert, S., Romeyke, M., Lenz, D., Erickson, H. M., Ananthakrishnan, R., Mitchell, D., Käs, J., Ulvick, S., and Bilby, C. (2005). Optical deformability as an inherent cell marker for testing malignant transformation and metastatic competence, *Biophys. J.*, **88**, pp. 3689–3689.
11. Ochalek, T., Nordt, F. J., Tullberg, K., and Burger, M. M. (1988). Correlation between cell deformability and metastatic potential in B16-F1 melanoma cell variants, *Cancer Res.*, **48**, pp. 5124–5128.
12. Lekka, M., Laidler, P., Gil, D., Lekki, J., Stachura, Z., and Hrynkiewicz, A. Z. (1999). Elasticity of normal and cancerous human bladder cells studied by scanning force microscopy, *Eur. Biophys. J.*, **28**, pp. 312–316.
13. Wu, H. W., Kuhn, T., and Moy, V. T. (1998). Mechanical properties of L929 cells measured by atomic force microscopy: Effects of anticytoskeletal drugs and membrane crosslinking, *Scanning*, **20**, pp. 389–397.
14. Rotsch, C., Braet, F., Wisse, E., and Radmacher, M. (2000). Drug-induced changes of cytoskeletal structure and mechanics in fibroblasts: An atomic force microscopy study, *Biophys. J.*, **78**, pp. 520–535.
15. Wakatsuki, T., Kolodney, M. S., Zahalak, G. I., and Elson, E. L. (2000). Cell mechanics studied by a reconstituted model tissue, *Biophys. J.*, **79**, pp. 2353–2368.
16. Goldmann, W. H., and Ezzel, R. M. (1996). Viscoelasticity in wild-type and vinculin deficient (5.51) mouse F9 embryonic carcinoma cells examined by atomic force microscopy and rheology, *Exp. Cell Res.*, **226**, pp. 234–237.
17. Radmacher, M. (1997). Measuring the elastic properties of biological samples with the atomic force microscope, *IEEE Med. Eng. Biol.*, **16**, pp. 47–57.
18. Sokolov, I. (2007). Atomic force microscopy in cancer cell research, in: *Cancer Nanotechnology* (Nalwa, H. S., and Webster, T., eds.), Chapter 1, American Scientific Publishers, New York, pp. 1–17.
19. Lekka, M., Gil, D., Pogoda, K., Dulińska-Litewka, J., Jach, R., Gostek, J., Klymenko, O., Prauzner-Bechcicki, Sz., Stachura, Z., Wiltowska-

Zuber, J., Okoń, K., and Laidler, P. (2012). Cancer cell detection in tissue sections using AFM, *Arch. Biochem. Biophys.*, **518**, pp. 151–156.

20. Lekka, M., Laidler, P., Ignacak, J., Łabędź, M., Lekki, J., Struszczyk, H., Stachura, Z., and Hrynkiewicz, A. Z. (2001). The effect of chitosan on stiffness and glycolytic activity of human bladder cells, *Biochim. Biophys. Acta (Mol. Cell Res.)*, **1540**, pp. 127–136.
21. Gumińska, M., and Ignacak, J. (1996). Electrophoretic pattern of cytosolic pyruvate kinase (PK) fractions A and B (Type L and M2) from normal rat liver and Morris hepatoma 7777, *Biochim. Biophys. Acta*, **1292**, pp. 99–105.
22. Glass–Marmor, L., and Beitner, R. (1997). Detachement of glycolitic enzymes from cytoskeleton of melanoma cells induced by calmodulin antagonist, *Europ. J. Pharmacol.*, **328**, pp. 241–248.
23. Levental, I., Georges, P. C., and Janmey, P. A. (2007). Soft biological materials and their impact on cell function, *Soft Matter*, **1**, pp. 229–306.
24. Baker, E. L., Bonnecaze, R. T., and Zaman, M. H. (2009). Extracellular matrix stiffness and architecture govern intracellular rheology in cancer, *Biophys. J.*, **97**, pp. 1013–1021.
25. Tang, X., Kuhlenschmidt, T. B., Zhou, J., Bell, P. Wang, F., Kuhlenschmidt, M. S., and Saif, T. A. (2010). Mechanical force affects expression of an in vitro metastasis-like phenotype in HCT-8 cells, *Biophys. J.*, **99**, pp. 2460–2469.
26. Baker, E. L., Lu, J., Yu, D., Bonnecaze, R. T., and Zaman, M. H. (2010). Cancer cell stiffness: Integrated role of three-dimensional matrix stiffness and transforming potential, *Biophys. J.*, **99**, pp. 2048–2057.
27. Prauzner-Bechcicki, Sz., Raczkowska, J., Madej, E., Pabijan, J., Lukes, J., Sepitka, J., Rysz, J., Awsiuk, K., Bernasik, A., Budkoswki, A., and Lekka, M. (2015). PDMS substrate stiffness affects the morphology and growth profiles of cancerous prostate and melanoma cells, *J. Mech. Behav. Biomed. Mater.*, **41**, pp. 13–22.
28. Pogoda, K., Chin, L., Georges, P. C., Byfield, F. J., Bucki, R., Kim, R., Weaver, M., Wells, R., Marcinkiewicz, C., and Janmey, P. A. (2014). Compression stiffening of brain and its effect on mechanosensitivity by glioma cells, *New J. Phys.*, **16**, pp. 075002.
29. Levental, I., Levental, K. R., Klein, E. A., Assoian, R., Miller, R. T., Wells, R. G., and Janmey, P. A. (2010). A simple indentation device for measuring micrometer-scale tissue stiffness, *J. Phys. Condens. Matter*, **22**, pp. 194120.

30. Xu, W., Mezencev, R., Wang, L., McDonald, J., and Sulchek, T. (2012). Cell stiffness is a biomarker of the metastatic potential of ovarian cancer cells, *PLoS ONE*, **7**, pp. e46609.

31. Mills, G. B., May, C., Hill, M., Campbell, S., Shaw, P., and Marks, A. (1990). Ascitic fluid from human ovarian cancer patients contains growth factors necessary for intraperitoneal growth of human ovarian adenocarcinoma cells, *J. Clin. Invest.*, **86**, pp. 851–855.

32. Lodish, H., Berk, A., Matsudaira, P., Kaiser, C. A., Krieger, M., Scott, M. P., Zipursky, S. L., and Darnell, J. (2004). *Molecular Cell Biology*, 5th ed. (W. H. Freeman and Company, New York, USA).

33. Carlier, M. F., and Pantaloni, D. (2007). Control of actin assembly dynamics in cell motility, *J. Biol. Chem.*, **282**, pp. 23005–23009.

34. Hinterdorfer, P., and Dufrene, Y. (2006). Detection and localization of single molecular recognition events using atomic force microscopy, *Nature*, **3**, pp. 347–355.

35. Zlatanova, J., and vanHolde, K. (2006). Single-molecule biology: What is it and how does it work?, *Mol. Cell*, **24**, pp. 317–2329.

36. Ikai, A. (2004). Nanomechanics of protein-based biostructures, *Jpn. J. Appl. Phys.*, **1**, pp. 7365–7375.

37. Oesterhelt, F., Oesterhelt, D., Pfeiffer, M., Engel, A., Gaub, H. E., and Muller, D. J. (2000). Unfolding pathways of individual bacteriorhodopsins, *Science*, **288**, pp. 143–146.

38. Afrin, R., Yamada, T., and Ikai, A. (2004). Analysis of force curves obtained on the live cell membrane using chemically modified AFM probes, *Ultramicroscopy*, **100**, pp. 187–195.

39. Lekenkari, P. P., and Horton, M. A. (1999). Single integrin molecule adhesion forces in intact cells measured by atomic force microscopy, *Biochem. Biophys. Res. Commun.*, **259**, pp. 645–650.

40. Laidler, P., Dulińska, J., Lekka, M., and Lekki, J. (2005). Expression of prostate specific membrane antigen in androgen–independent prostate cancer cell line PC–3, *Arch. Biochem. Biophys.*, **435**, pp. 1–14.

41. Lekka, M., Laidler, P., Łabędź, M., Kulik, A. J., Lekki, J., Zając, W., and Stachura, Z. (2006). Specific detection of glycans on a plasma membrane of living cells using atomic force microscopy, *Chem. Biol.*, **13**, pp. 505–512.

42. Lekka, M., Gil, D., Dąbroś, W., Jaczewska, J., Kulik, A. J., Lekki, J., Stachura, Z., Stachura, J., and Laidler, P. (2011). Characterization of N-cadherin unbinding properties in non-malignant (HCV29) and malignant (T24) bladder cells, *J. Mol. Recognit.*, **24**, pp. 833–842.

43. Li, F., Redick, S. D., Erickson, H. P., and Moy, V. T. (2003). Force measurements of the $\alpha5\beta2$ integrin-fibronectin interaction, *Biophys. J.*, **84**, pp. 1252–1262.

44. Zhang, X., Craig, S. E., Kirby, H., Humphries, M. J., and Moy, V. T. (2004). Molecular basis for the dynamic strength of the integrin $\alpha4\beta1$/VCAM-1 interaction, *Biophys. J.*, **87**, pp. 3470–3478.

45. Taubenberger, A., Cisneros, D. A., Friedrichs, J., Puech, P. H., Muller, D. J., and Franz, C. M. (2007). Revealing early steps of $\alpha2\beta1$ integrin-mediated adhesion to collagen type I by using single-cell force spectroscopy, *Mol. Cell Biol.*, **18** pp. 1634–1644.

Chapter 2

Cell Structure and Functions

A cell is a basic unit forming all living organisms that is capable to maintain various, independent functional processes [1]. The simplest example is a bacterium where all cellular processes are carried out within a single cell body. In multicellular organisms, different kinds of cells perform different functions. This enables to create highly specialized tissues (connective, muscle, nervous, and epithelial ones) being the basis for organs formation. Despite that, the basic cellular structure remains similar in all organs (Fig. 2.1). A central part of a cell is a nucleus, which is a membrane-enclosed organelle operating as a storage place for genetic information encoded in DNA strands. Close to cell nucleus, an endoplasmic reticulum with associated ribosomes is located. This organellum is responsible for a protein and lipid synthesis. Newly synthesised proteins and lipids are sorted in the Golgi apparatus, from which they are distributed to other cellular compartments or membranes. The mitochondria are organelles where energy is stored. They contain two major membranes, i.e., the outer and the inner one. The inner membrane has much more restricted permeability and it is loaded with proteins involved in electron transport and ATP (adenosine triphosphate) synthesis, used for energy production. The outer membrane has many protein-based pores that enable transport of ions and small molecules. The lysosomes are specialized organelles that function as the

Cellular Analysis by Atomic Force Microscopy
Malgorzata Lekka

ISBN 978-981-4669-67-2 (Hardcover), 978-1-315-36480-3 (eBook)
www.panstanford.com

digestive system inside cells and are responsible for degradation of a material taken up from outside the cell and for digestion of obsolete cellular components. Lysosomes contain arrays of enzymes that are capable to break down any type of biological polymers—proteins, nucleic acids, carbohydrates, and lipids. Within a cellular space, there are multiple types of various vesicles (e.g., endosomes) that are required for the molecular transport within the cell and between the cell and its environment.

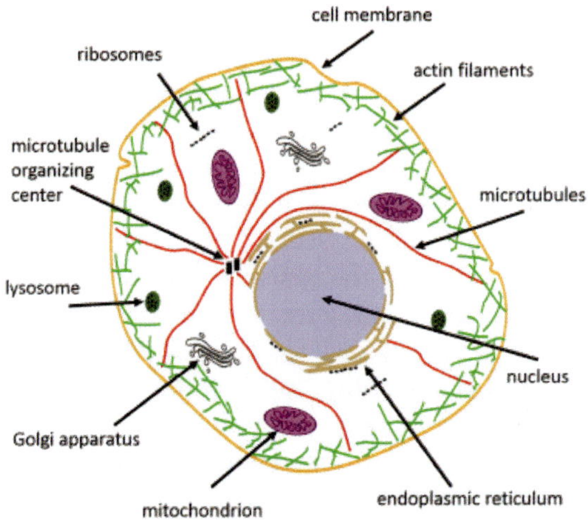

Figure 2.1 Schematic representation of a cell showing its basic structures.

Each cell is embedded within an external environment, i.e., *extracellular matrix* (*ECM*) composed of various constituents such as proteins, proteoglycans, hyaluronan, etc. The border between the cell and the ECM constitutes a plasma membrane, which acts as a selective molecular filter and regulates the communication between cells. All intracellular organelles are embedded in a cytoplasm, which fills the cell interior. *The cytoplasm* contains two fractions, i.e., a *cytosol* (a liquid fraction) and a *cytoskeleton* (a network of protein filaments).

Cytosol is the intracellular fluid comprises of water, dissolved ions, large water-soluble molecules, smaller molecules, and proteins. It is not a simple solution of molecules—it should be considered as a highly concentrated medium with complex composition and gel-like consistency. Within a cytosol, multiple levels of organization can be found. These include concentration gradients of small molecules such as calcium, large complexes of enzymes that act together to carry out metabolic pathways, and protein complexes such as proteasomes that enclose and separate parts of the cytosol.

Cytoskeleton is a mesh-like structure composed of various filamentous proteins that forms a cellular scaffold embedded within a cell cytoplasm. Apart of its structural functions related to maintaining cellular shape and providing the tool for organelles arrangements, the cytoskeleton participates in various processes through interactions with other proteins, such as cell migration, muscle contraction, cell division, cell adhesion, and intracellular transport. The cytoskeleton helps to establish regularity within the cytoplasm and, together with the plasma membrane, determines the mechanical stability of the cell.

Before considerations on the use of the atomic force microscopy to analyze cellular properties, the essential information on the cell's milieu and structure of essential cellular components is provided below.

2.1 Extracellular Matrix

In each animal tissue, cells are surrounded by an extracellular matrix (ECM), which provides a physical scaffold for cells and initiates biochemical and biomechanical processes [2]. Its importance has been already shown in diseases with genetic ECM abnormalities such as various skin diseases, chondrodysplasias, or Ehlers–Danlos and Marfan syndromes [3].

The ECM is a mixture of many different components including water, proteins, polysaccharides, and proteoglycans. The composition varies between tissues despite of the same building elements. The internal structure of the matrix is well organized, which comes from both intrinsic properties of building molecules and activities of cells. The ECM plays an important role in regulating cell functions such as cellular morphology, adhesion, migration, proliferation, apoptosis, etc. The active participation of the ECM in cellular functions denotes that the molecular composition of the matrix is not constant but undergoes constant remodeling [4, 5].

2.1.1 The ECM Proteins

There exist a huge number of proteins constituting the major component of the ECM, including collagens, elastin, fibrillins, fibronectins, laminins, and nidogens.

The mostly abundant proteins of the ECM are *collagens* forming a large family of at least 19 isoforms (Fig. 2.2) [6].

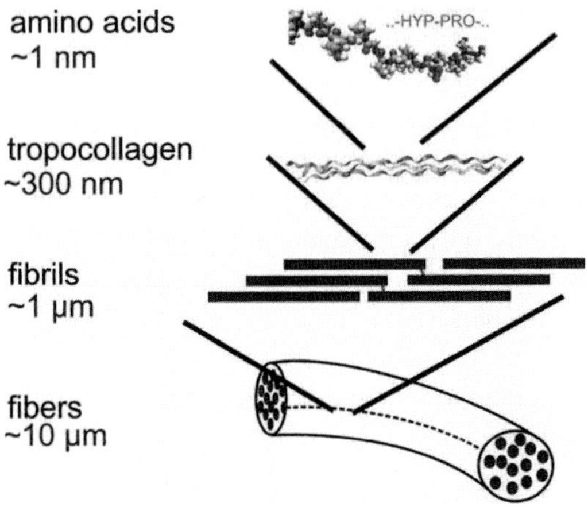

Figure 2.2 Schematic view of the hierarchical structure of collagen fibers, ranging from the amino acid sequence to collagen fibers with lengths of the order of 10 microns. Reprinted with permission from [6].

All collagen molecules have a tight-handed helix of three individual collagen chains (called *tropocollagen*). Each individual collagen chain has a left-handed helical structure. Collagen types I, II, and III account for around 90% of total quantity of collagens in the organism. These molecules form strong fibrils that can include small amount of the other fibrillar collagens like V or XI. The mechanical properties of collagen fibers can be changed by the presence of collagen type IX, XII, and XIV. The latter types are commonly present in connective tissues. The basal membranes of the epithelium contain collagen IV that forms 3D gels [7].

The other class of the ECM proteins are *fibronectins* [8, 9]. Each fibronectin molecule is a dimer composed of two long monomers characterized by a modular structure, linked to each other through disulfide bonds located near their carboxyl ends (Fig. 2.3).

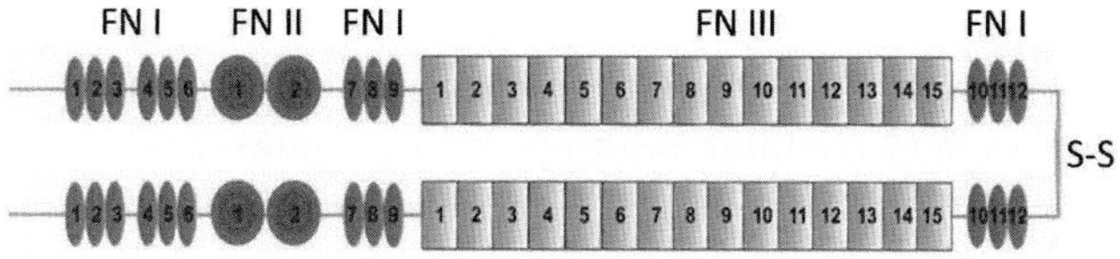

Figure 2.3 Illustration of fibronectin dimer. It is composed of two monomers linked by disulfide bonds (S-S). Each monomer consists of 12 domains of type I, 2 of type II and 15 of type III.

The fibronectin monomer is composed of about 2300 amino acids, which corresponds to the length of about 870 nm, estimated as a product of the number of amino acids multiplied by the length of the monomer unit (0.38 nm). In the tertiary structure of fibronectin three types of domains can be distinguished [8]. There are 12 domains of type I (FN I), two domains of type II (FN II) and about 15 domains of type III (FN III). FN domains have the length of approximately 16 nm, 24 nm and 34 nm, corresponding to 44 (FN I), 63 (FN II), and 90 (FN III) amino acids residues. Each domain has a β-sheet structure maintained mainly by hydrophobic interactions [8] and by hydrogen bonds [9]. A single fibronectin molecule possesses various binding sites for heparin sulfate proteoglycans, integrins, and collagens [9].

The third most studied family of the ECM proteins are *laminins* [10]. They have cross-like shape composed of three chains (α, β, and γ) that are assembled into a coiled-coil structure forming a long arm (Fig. 2.4).

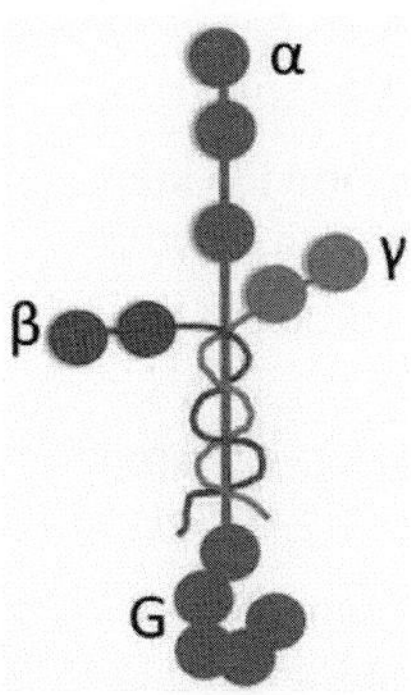

Figure 2.4 Laminins are cross-shaped proteins with three arms ($\alpha\beta\gamma$ chains) that are brought together forming a coiled-coil structure terminated with a globular domain (G) at the end of a long α-chain.

In mammals, five α, three β, and three γ chains have been identified, bringing a final combination of at least 15 isoforms [10]. Each chain contains multiple 7 long amino acid sequences that enable them to come together. The structure of the coiled coil maximizes the number of non-covalent bonds formed between the subunits and provides the structural stability. When binding to integrin, the C-end of the laminin chains is indirectly involved within the G domain.

2.1.2 Proteoglycans

Proteoglycans are glycosylated proteins that are composed of a protein with one or more covalently attached glycosaminoglycan (GAG) [11]. Usually, GAGs (heparan or chondroitin sulfate) are attached through a tetrasaccharide bridge to serine (Ser) while keratan sulfate is attached to asparagine (Asp). Proteoglycans follow a simple rule: *protein* + *GAGs* = *proteoglycan*. In nature, there are few proteoglycans sharing a simple structure, which differs them from glycoproteins. The GAGs chains form long, non-branched, linear carbohydrate polymers that are negatively charged under physiological conditions due to the occurrence of sulfate and uronic acid groups [11]. Proteoglycans are usually categorized depending on a type of glycosaminoglycan chains and molecular mass (Table 2.1).

Table 2.1 Exemplary representative of the most common proteoglycans

Glycosaminoglycans	Small proteoglycans (M_w < 100 kDa)	Large proteoglycans (M_w > 100 kDa)
Chondroitin sulfate /dermatan sulfate	Decorin (36 kDa) Biglycan (38 kDa)	Versican (260–370 kDa)
Heparin sulfate /chondroitin sulfate	Testican (44 kDa)	Perlecan (400–470 kDa)
Chondroitin sulfate	Bikunin (25 kDa)	Neurocan (136 kDa) Aggrecan (220 kD)
Keratin sulfate	Fibromodulin (42 kDa) Lumican (38 kDa)	

Apart from ECM proteins, proteins, proteoglycans are major components of the animal extracellular matrix that fill the space between cells in an organism. They form large complexes linked to other proteoglycans or to hyaluronic acid or to fibrous matrix proteins (such as fibronectin, laminin or collagen). Proteoglycans bind cations (such as sodium, potassium and calcium), water, and regulate the movement of molecules through the extracellular matrix. There are evidences showing their influence on the activity and stability of proteins and also on signaling molecules present within the matrix [1, 11].

2.1.3 Other Components of the ECM—Hyaluronan

Among other components of the ECM, *hyaluronic acid* (HA, also so-called hyaluronan) should be mentioned (Fig. 2.5) since its negatively charged polysaccharide molecules produce hydrated gels filling the ECM.

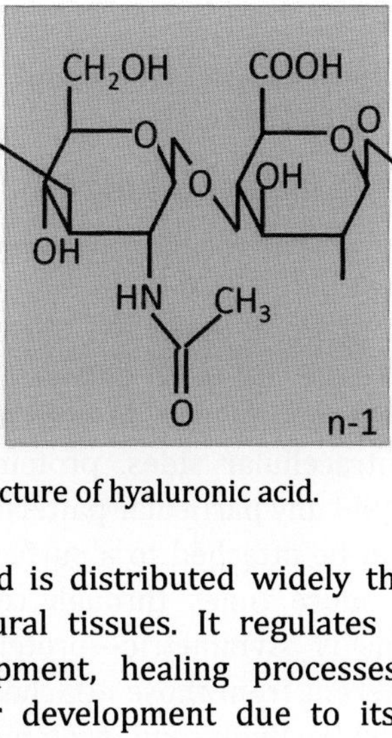

Figure 2.5 The structure of hyaluronic acid.

Hyaluronic acid is distributed widely throughout connective, epithelial, and neural tissues. It regulates cell behavior during embryonic development, healing processes, inflammation, and also during tumor development due to its interactions with a specific transmembrane receptor, CD44 [1, 12].

2.2 Cell Membrane

The external boundary of a cell is provided by *a membrane* [1]. It is not only a passive barrier separating a cell from an environment but it actively participates in various phenomena needed for cell functioning. Membranes provide a tool to maintain cell integrity and also they divide internal cellular space into compartments, where, sometimes, quite contradictory processes occur. The biological activity of cell membrane are a result of its structure and exclusive physical properties. Cell membrane is highly flexible, which enables relatively fast changes in cellular shape observed, for example, during cell division or cell migration. Due to selective permeability, cellular membrane regulates the transport of certain molecules and ions to cell interior/exterior, within a cell, and also within a specific cellular compartment. Cell membrane contains a variety of molecules and proteins participating in various cellular processes. At the cell surface, some proteins are responsible for the cell-to-extracellular matrix and for cell-to-cell interactions while the other ones move specific organic solutes and inorganic ions across the membrane. When situated inside the cell, various membranes take part in such processes as lipid synthesis or energy transduction in mitochondria [1].

2.2.1 Membrane Structure

The membrane is a sheet-like structure with thickness of about 6–10 nm, mainly composed of lipids [1]. They are relatively small molecules having both hydrophilic and hydrophobic parts. Hydrophobic regions of lipid molecules face each other while their hydrophilic parts are exposed outward. This leads to the formation of a so-called *lipid bilayer*. In such a bilayer, regardless of extracellular or intracellular sides, proteins are attached in a random fashion without any particular pattern (Fig. 2.6).

Proteins either can be attached to a surface of a lipid bilayer or can span one or more times through cell membrane. The orientation of proteins is asymmetric—proteins on one side of a lipid bilayer are different from those attached to the other side. On the extracellular side, lipids and proteins are glycosylated, i.e., they possess covalently attached oligosaccharide structures called *glycans*. Inside the cell, proteins are frequently linked to cytoskeleton facing the cytosol. It is important to underline that both proteins and lipids are free to laterally move in a plain of a bilayer but movement from one bilayer side to another is strongly restricted.

All membranes have similar basic structure and certain common functions; however, each type of membrane present within a cell interior plays certain distinctive role due to a unique set of proteins and lipid composition. For example, neurons possesses so-called *myelin sheath*, which is a plasma membrane wrapped several times around a cell. The myelin sheath acts as a passive electric insulator because of its high content of lipids. Distinct composition of membranes in mitochondria is crucial for enzyme-catalyzed processes, and thus, these membranes have more proteins than lipids [1].

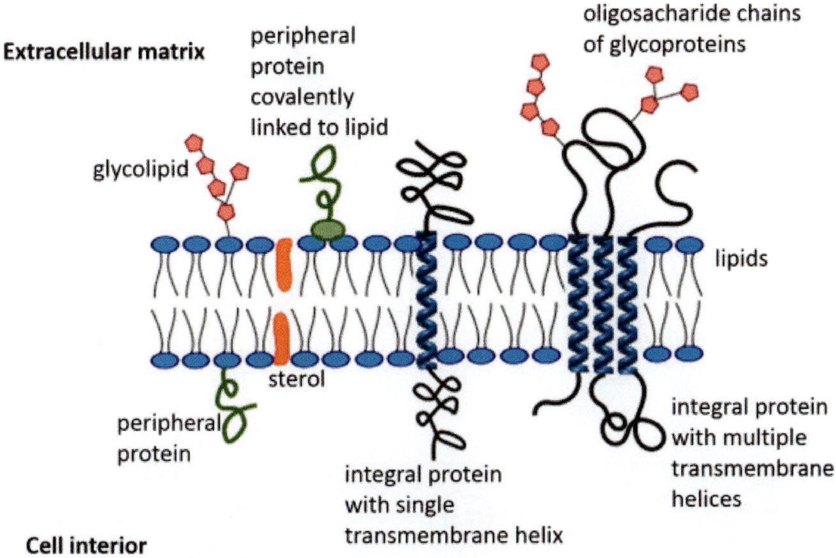

Figure 2.6 Scheme of a fluid mosaic model for cell membrane.

2.2.1.1 Lipids

Lipids are a large group of organic compounds of various composition and structure. A lipid molecule always contains a long chain composed of fatty acids providing a poor solubility in water, which is related to the presence of a large number of hydrophobic nonpolar hydrocarbon groups ($-CH_2-$). Besides to the hydrocarbon chains, lipid molecules comprise hydrophilic polar groups (i.e., ester, phosphate, hydroxy, or amino ones), which are capable to interact with water by the formation of hydrogen bonds. Due to their amphipathic (i.e., partly hydrophobic and partly hydrophilic) properties, lipids are capable to form a lipid bilayer where the hydrophobic effect and van der Waals interactions cause that the polar head groups are oriented towards water (outside of the bilayer) while hydrophobic tails are hidden [1, 2]. The lipid bilayer of most animal membranes is composed mainly of three classes of molecules: phospholipids, glycolipids, and steroids (Fig. 2.7).

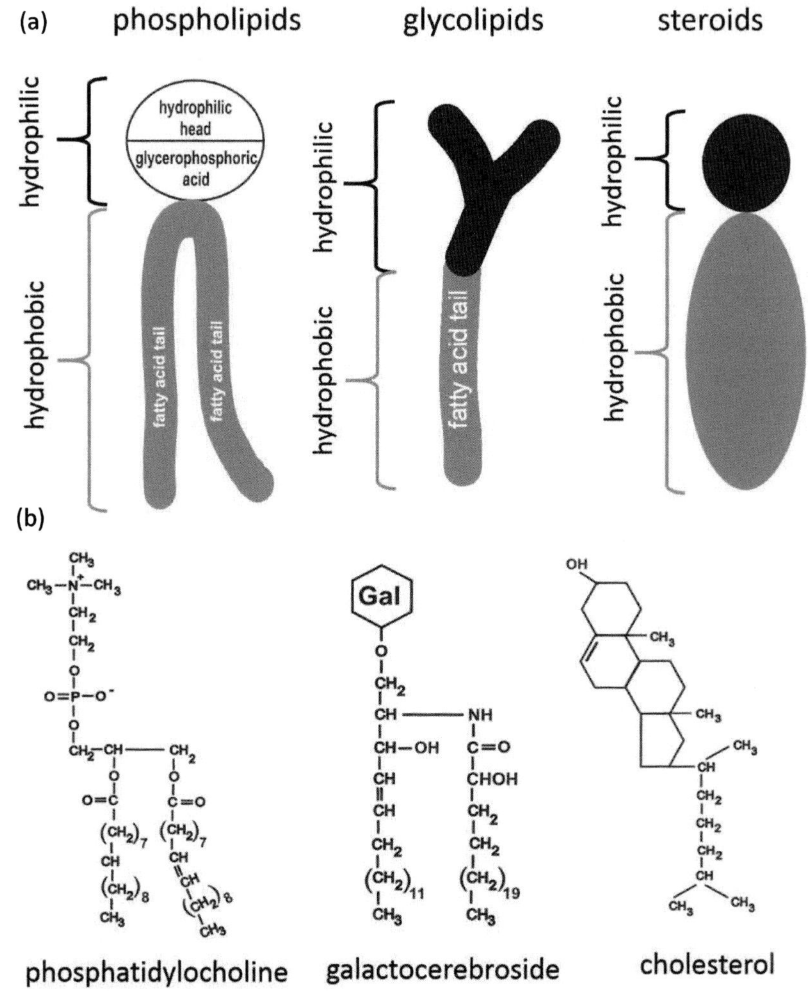

Figure 2.7 (a) Schematic structures of the three most common groups of lipids, i.e., phospholipids, glycolipids, and steroids. (b) Chemical structure of one molecule representing three classes of membrane lipids (phosphatidylocholine, cholesterol, and galactocerebroside).

Phospholipids represent a large group of lipids, whose characteristic feature is the presence of phosphate group in the hydrophilic part of the molecules [1, 2]. The hydrophobic part consists of two fatty acid chains* (Fig. 2.7a). The differences in structure of various phospholipids exist in both polar and hydrophobic parts. The former can have either ionic or amphipathic character, whereas hydrophobic tails can contain hydrocarbon chains of different lengths and degrees of unsaturation [2]. Within phospholipids, two sub-families of molecules can be distinguished, namely *phosphoglycerides and sphingolipids.*

Phosphoglycerides are derivatives of glycerol-3-phosphate. They have two hydrocarbon chains in the glycerol backbone and one phosphoric acid (or one phosphoric acid esterified group). The main members of phosphoglycerides are phosphatic acid (PA), phosphatidylocholine (PC), phosphatidyl-ethanolamine (PE), phosphatidylserine (PS, Fig. 2.7b), and phosphatidylinositol (PI). Hydrophilic regions of phospholipid molecules contain acidic phosphate groups possessing a negative charge (at pH 7.0). The other component of the hydrophilic region may be serine, choline, ethanolamine, or inositol. Both choline and ethanolamine contain the positively charged amino groups. The presence of two oppositely charged groups in the polar head of a lipid gives it the character of a dipolar ion with no net charge. Phospholipids containing serine or inositol have a negative net charge. *Sphingolipids* are derivatives of sphingosine, which is a long unsaturated alcohol chain. The most abundant, representative molecule is a sphingomyelin residing mostly in nerve cells but also present in kidney tissue and blood. This molecule contains very little amount of unsaturated fatty acids attached with sphingosine through nitrogen atom.

Glycolipids form the second group of lipids. They have covalently bound mono- or polysaccharides in the hydrophilic region (through glycosidic bonds). The most common saccharides molecules are galactose, glucose or lactose [1, 2]. Glycolipids can be found in lipid membranes of various organs like brain, nerve system, spleen, kidney, lung, liver, and erythrocytes. Analogously as phospholipids, glycolipids have amphipathic character (Fig. 2.7a). There are several sub-families of glycolipids such as glycosphingo-

*In the case of lysophospholipids, there is only one fatty acid chain.

lipids, commonly found in nerve tissue, made up of ceramides connected to a carbohydrate moiety. Galactocerebroside (Fig. 2.7b) is a type of glycolipid that is commonly found in the myelin sheath around the nerves of vertebrates.

The third main groups of cell membrane lipids are *steroids* (Fig. 2.7a). The main feature of these molecules is the ring system of three cyclohexanes and one cyclopentane [2]. There exist a variety of functional groups that may be attached. Like other membrane lipids, they are amphipathic. They contain the hydrophilic head (-OH) that is directly linked with hydrophobic part of the molecule. The most common molecule is cholesterol, in which a hydroxyl group functions as polar head group (Fig. 2.7b). Cholesterol is a ring compound containing a branched side chain. In the body, it is formed from two-carbon elements (acetic acid radical) and it is a raw material for the production of bile acids and various hormones.

Lipid composition of various cell membranes varies among distinct structures within a cell, depending on functions performed by specialized organelles [13]. The Fig. 2.8 shows the synthesis site of lipids that are involved in signaling and recognition pathways.

The major glycerophospholipids assembled in the endoplasmic reticulum (ER) are phosphatidylcholine, phosphatidylethanolamine, phosphatidylinositol, phosphatidylserine, and phosphatidic acid. In addition, the ER synthesizes ceramides, galactosylceramide, cholesterol, and ergosterol. Both the ER and lipid droplets participate in steryl ester and triacylglycerol synthesis. The Golgi lumen is the site of synthesis of sphingomyelin, complex glycosphingolipids, and yeast inositol sphingolipid. Phosphatidylcholine is also synthesized in the Golgi, and may be coupled to protein secretion at the level of its diacylglycerol precursor. Approximately 45% of the phospholipids (mostly phosphatidylethanolamine, phosphatidic acid, and cardiolipin) in mitochondria are autonomously synthesized by this organelle. The inner membrane of late endosomes produces mostly bis(mono-acylglycero) phosphate (BMP).

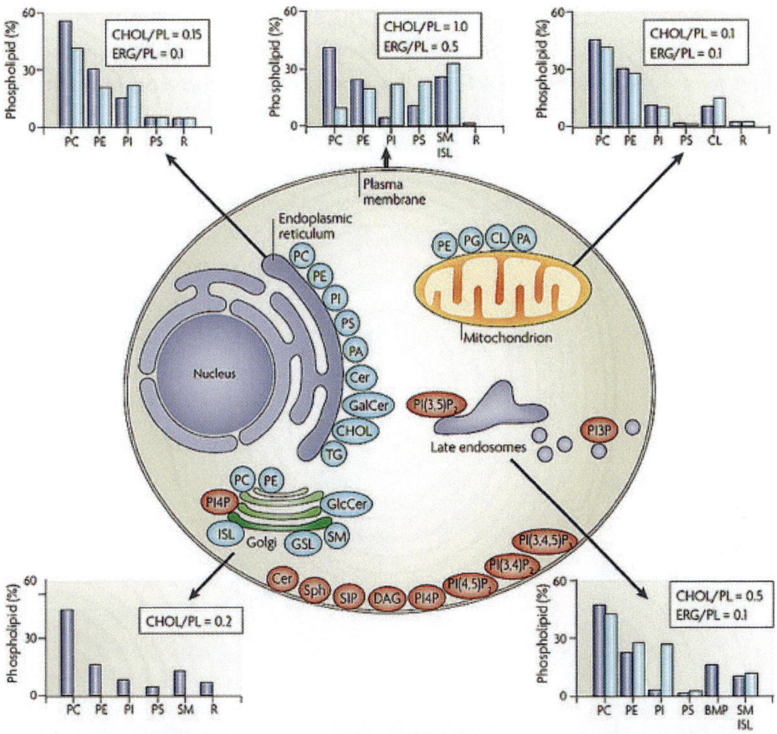

Figure 2.8 Lipid composition of various cell membranes expressed as a percentage of the total amount of phospholipids (PL) in mammals (blue) and in yeast (light blue). As a measure of sterol content, the molar ratio of cholesterol (CHOL in mammals) and estrosterol (ERG in yeast) to phospholipid is also included. The synthesis sites of a particular lipids shows various organelles involved in this process (PA—phosphatidic acid, PC—phosphatidylcholine, PE—phosphatidylethanolamine, PI—phosphatidylinositol, PS—phosphatidylserine, PG—phosphatidylglycerol, CL—cardiolipin, GalCer—galactosylceramide, TG—triacylglycerol, SM—sphingomyelin, DAG—diacylglycerol, GSLs—glycosphingolipids, and R—remaining lipids). Reprinted with permission from [13].

2.2.1.2 Proteins

As the lipid bilayer physically separates the interior of a cell from its surrounding, the transport through membrane is realized by other molecules. These are the differently specialized proteins that serve as pumps, carriers, receptors, etc., enabling a more or less selective recognition of specific compounds and/or their translocation in and out the cell [1, 2].

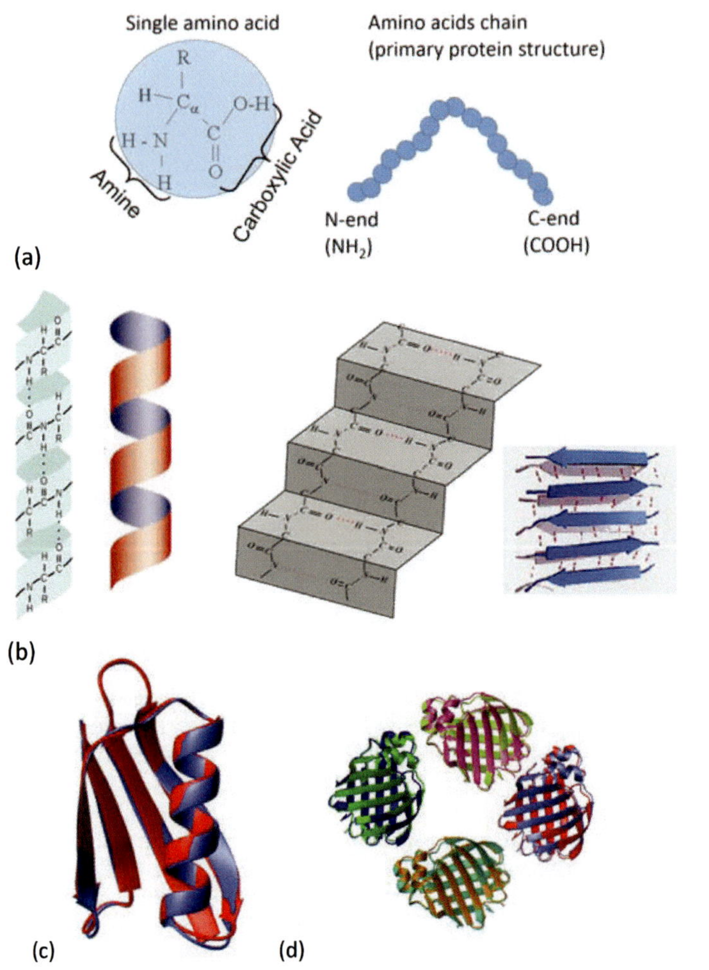

Figure 2.9 Four levels of organizations in protein structure: (a) primary (a single chain of amino acids), (b) secondary, namely, α-helix and β-sheets, (c) tertiary (3D structure of polypeptide chains), and (d) quaternary ones showing a spatial organization of proteins composed of more than one polypeptide chain. Prepared based on [1] and hypothetical tertiary protein structure.

There are only 20 building blocks of proteins (called amino acid) that are used to produce a huge number of proteins with distinct biological functions. Such diversity stems from proteins structure systematized into four levels of the organization:

(1) *primary protein structure*, which is the sequence of amino acids linked in a single polypeptide chain (Fig. 2.9a);
(2) *secondary protein structure* stemming from the regular, repetitive amino acid orientation in the protein chain as a result of the local organization of the hydrogen bonds. There are two typical structures: *α-helix* and *β-strand* (Fig. 2.9b);
(3) *tertiary protein structure*, which is a three-dimensional structure of the whole polypeptide chain being the result of distinct interactions like ionic, hydrogen, van der Waals forces occurring between the side groups of various amino acids and the presence of two bridges sulfide formed between a pair of adjacent cysteine residues (Fig. 2.9c);
(4) *quaternary protein structure* related to the spatial configuration of protein built from more than one polypeptide chain (Fig. 2.9d).

There are many types of membrane proteins that can be roughly divided into three main classes: integral, peripheral, and lipid-anchored proteins. The integral proteins are fully incorporated into the cell membrane and contact both the inside and the outside of the cell. The peripheral proteins are non-covalently bound to the polar head group of the lipid bilayer or to an integral membrane protein. The lipid-anchored proteins are membrane proteins covalently bound to either a phospholipid or a fatty acid that is embedded in the lipid bilayer [1, 2].

2.3 Surface Receptors

The interactions between cells or between a cell and ECM require presence of molecules on a cell surface (called *surface receptors*). A particular group of surface receptors participating in adhesion phenomena is called *adhesion molecules.* The interaction of surface receptors with given ligands initiates a complex cascade of events inside the cell, influencing cellular behaviors in different ways [1].

All adhesion molecules are present as integral plasma membrane proteins, and they are exposed to the extracellular environment. They are classified into four main families namely, cadherins, integrins, selectins, and immunoglobulin-like superfamily. The interactions between these molecules can have either homophilic (with the same partner) and heterophilic (a complementary, distinct ligand is needed) character.

Adhesion molecules are variously expressed in different cells of the same organism and often by the same cell at various stages of their development since they mediate numerous distinct interactions (contacts) between cells and their surroundings, i.e., between other cells and/or with proteins embedded in the extracellular matrix. The structural arrangement of receptor molecules defines, among others, such parameters such as strength, lifetime, association, and/or dissociation constants characteristic for specifically recognized ligands, antibodies, and other proteins.

It should be mentioned that adhesion of cells, frequently, proceeds through clusters containing a large number of adhesion receptors. They mediate contact to a group of specific ligands either carried by other cells or present in the extracellular matrix. These clusters are assemblies of transmembrane proteins that are held in the plasma membrane of the cell by hydrophobic interaction. On the cytoplasmic side (i.e., inside the cell), they can be linked to the actin cytoskeleton. A binding pocket on the extracellular side allows specific interaction with appropriate ligands.

2.3.1 Integrins

The most diverse group of adhesion molecules are *integrins* that are heterodimers composed of α and β subunits (Fig. 2.10).

There are at least 18 well-recognized α-type and 9 β-type integrin subunits [1, 2]. Their extracellular domains participate in bivalent metal ion-dependent interactions with various extracellular matrix proteins and other receptors of neighboring cells belonging predominantly to the immunoglobulin superfamily. It is worth to mention that metal ions play a critical role in the ligand binding function of all integrin heterodimers. It has been shown that the presence of arginine-aspartate-glycine (RDG) amino acid sequence has been ascertained for the interaction with ECM proteins such as fibronectin, vitronectin or collagen I. Upon binding to their ligands, integrins activate members of intracellular signaling pathway cascades, transducing in and out signals that can stimulate or regulate motility and invasiveness, cell growth, and survival. The expression of integrins depends on the tissue of origin and the degree of differentiation. Cancer cells very often switch the types of expressed integrins, favoring those that transmit growth signals [1, 14].

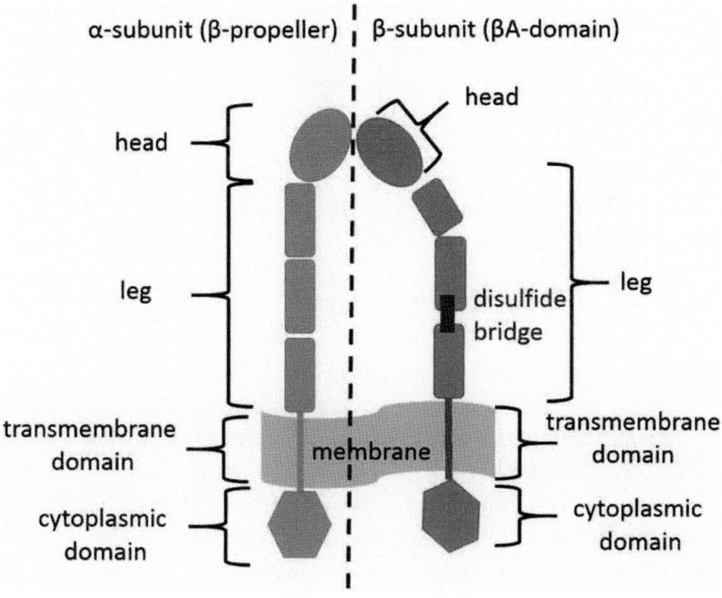

Figure 2.10 The integrin structure. Integrins are heterodimeric receptors consisting of α- and β-subunits (the β-propeller and the βA domain, respectively). They assemble by non-covalent interactions to form a *head* that provides a ligand-binding site. The head comprises two long *legs* ended up with transmembrane and cytoplasmic domains.

2.3.2 Cadherins

Cadherins are the most important adhesion molecules that are involved in the formation of cell–cell adherent junctions. They are critical for the maintaining of tissue architecture, cell polarity, limiting cell movement, proliferation, and also in the transformation and invasion of cancer cells [1, 15]. Cadherins are transmembrane proteins having both extracellular and cytoplasmic domains (Fig. 2.11).

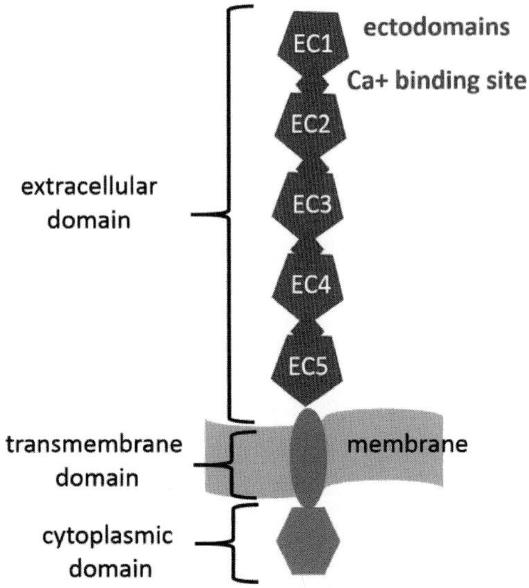

Figure 2.11 The scheme showing a structure of classical cadherins.

The extracellular domain of cadherins consists of five repeats (called *ectodomains*), each of about 110 amino acids. Between two repeats, calcium ions are bound, participating in the formation of calcium-dependent, homophilic bonds. The cytosolic domain is directly associated with α-catenin or/and plakoglobin (γ-catenin). These proteins are members of armadillo family, sharing a common motif called "*armadillo repeat domain.*" Both α-catenin and plakoglobin are composed of 12 such repeats (40 to 42 amino acids residues) with unique amino- and carboxyl-terminal domains. They share 76% identity in the armadillo repeats, and thus, they are highly homologous and often interact with the same partners [16, 17]. However, there is one exception: plakoglobin can associate with both classical (e.g. E- or N-cadherin) and desmosomal cadherins, while α-catenin associates only with the members of the classical cadherin family. Both α-catenin and plakoglobin bind to β-catenin, which links the cadherin/catenin complex to the actin cytoskeleton [18]. Cadherins are divided into a few subgroups (Fig. 2.12).

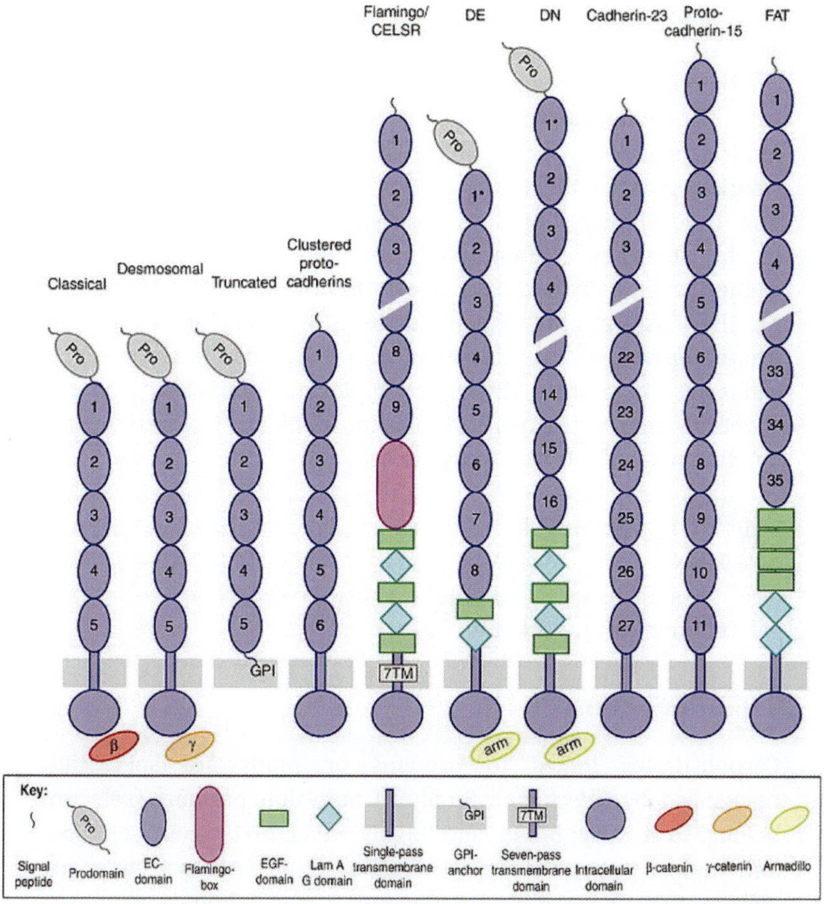

Figure 2.12 Schematic representation of members of the cadherin family, which share a common structural motif, i.e., extracellular domains called *ectodomains* (*re-printed with permission from [19]*).

In both the extracellular and the cytoplasmic domains, each class of cadherin family possesses its own characteristics [19]. Classical cadherins contain extracellular calcium binding domains, including the histidine-alanine-valine (HAV) and tryptophan (W) conserved motifs, along with cytoplasmic domain that binds to β-catenin and protein p120 and connects the cadherin to the actin cytoskeleton. The most common members of classical cadherins are *E-cadherin*, which is mostly expressed in epithelial tissue of various organs and recognized as one of the suppressors of cancers, *N-cadherin* that is mainly found in neural tissue and many cells at their fetal stage of development, but also it is common to many metastatic tumors and therefore known as a tumor inducer, and *P-cadherin*, usually found initially in the placenta. Another class of the cadherin family are *desmosomal cadherins* that have similar motifs as classical ones, but instead of one tryptophan, they contain two tryptophan molecules, truncated cadherins that do not have a cytoplasmic domain but, instead of, they use glycosyl-phophatidylinositol (GPI) as an anchor, *protocadherins* with extracellular domain similar to desmosomal cadherins but with distinct cytoplasmic domains, and *fat cadherins* that with their cytoplasmic domains interacting with a different set of cytoplasmic proteins.

Cadherins have been found to maintain calcium-dependent cell–cell recognition and adhesion. They are involved in such processes as morphogenesis, mechanotransduction, cell signaling, physical homeostasis, and others. The altered cadherin expression is a common feature of metastasis.

2.3.3 Selectins

Carbohydrates linked to proteins or lipids are specifically recognized by *selectins*, a family of adhesion proteins (selectins belong to a very broad and highly diversified group of plant and animal proteins called *lectins* [20]). All selectins are glycoproteins with a Ca^{+2}-dependent carbohydrate-binding site located at the extracellular part, followed by a single epidermal growth factor domain, a variable number of regulatory domains, a single transmembrane polypeptide, and a fairly short cytoplasmic domain (Fig. 2.13).

Selectins mediate the cell–cell contacts by binding *via* their lectin domain specific to a carbohydrate-containing counter-receptor on target cells. The recognized members of the selectin family are L-selectins (constitutively expressed on leukocytes), P-selectins (mainly expressed on platelets but also on endothelial cells), and E-selectins (inducible endothelial cell protein). They bind to selectively to particular oligosaccharides such as sialyl-Lewis x (sLex) attached to glycoproteins and glycolipids on most leukocytes and some endothelial cells. The results of numerous studies strongly suggest that one of the key factors in metastasis is the presence on tumor cells of a high-density sLex structure, and ligands of E- and P-selectins. This complex recognition is followed by the aggregation of platelet and tumor cells leading probably to their attachment to the endothelium, extravasation, and colonization of a tissue.

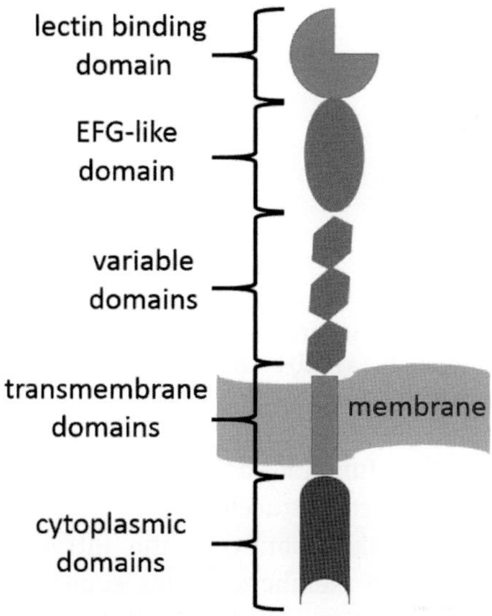

Figure 2.13 The structure of selectin molecule. They are classified basing on similarities in specificity of their carbohydrate recognition domains.

2.3.4 Immunoglobulin Family

Among the adhesion molecules there are proteins belonging to the *immunoglobulin superfamily* [21]. Its members (also called immunoglobulin-like cell adhesion molecules, ICAMs) possess a common structural motif, an immunoglobulin fold composed of about 70–110 amino acids that form two anti-parallel β-plated sheets stabilized by disulphide bonds [1, 21]. Structurally, these Ig superfamily share a composition of constant/variable domains with disulfide bonds followed by a transmembrane domain (Fig. 2.14).

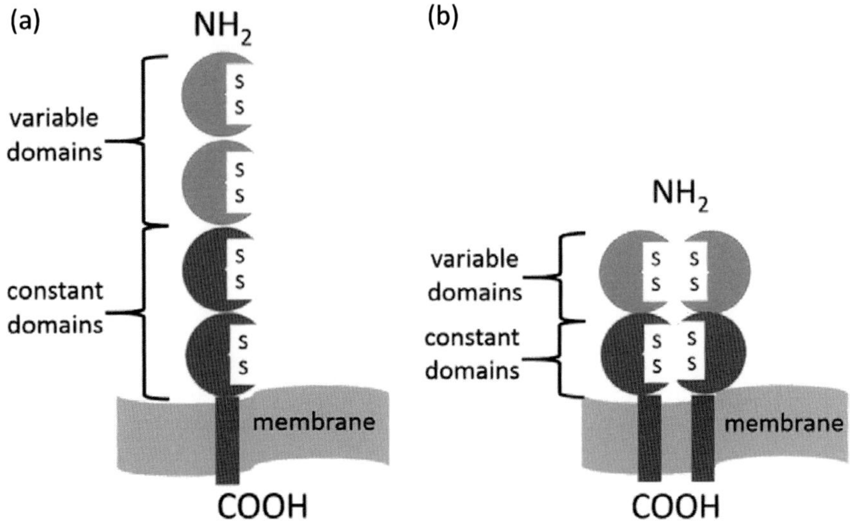

Figure 2.14 Schematic structures of two exemplary members of Ig superfamily: (a) CD4 co-receptor, and (b) T-cell antigenic receptor.

Immunoglobulins function as cell adhesion and signaling receptors that transduce extracellular signals from neighboring cells or the extracellular matrix to the intracellular signaling machinery. Most of the members of this superfamily participate in the cell–cell recognition, immunological processes, and also in cancer metastasis.

2.3.5 Glycans

Many of the integral membrane proteins, on the extracellular side of the membrane, contain one (rarely) or more oligosaccharide components (due to the origin also called carbohydrate moieties). An individual oligosaccharide chain, covalently bound to a protein, is called a *glycan* and such proteins are called *glycoproteins*.

Glycans are complex carbohydrates composed of single sugar units that are usually referred to the carbohydrate moiety attached either to proteins or to lipids [22]. Depending on the bond structure, there are two types of glycans: O- and N-linked ones (Fig. 2.15).

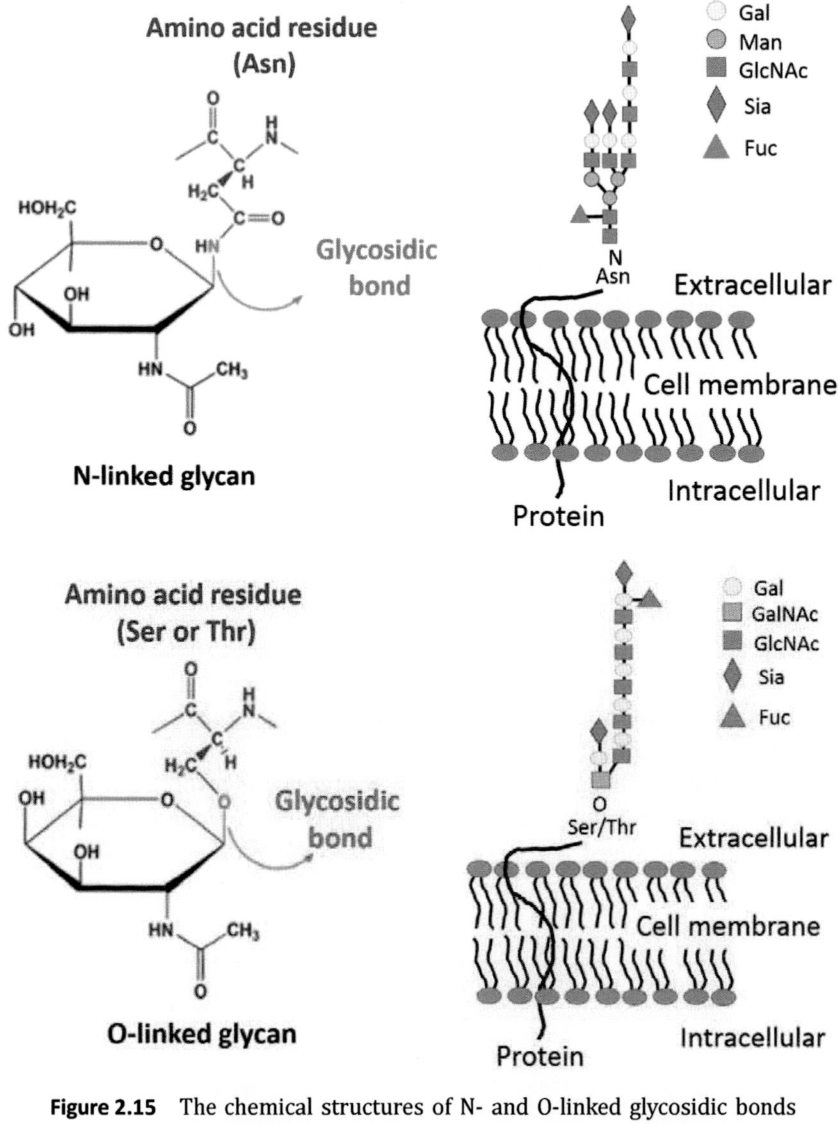

Figure 2.15 The chemical structures of N- and O-linked glycosidic bonds accompanied with the schematic view of resulting N- and O-linked glycans. Specific sugars are usually represented by colored geometric symbols as presented in the legend (Gal–galactose; Man–mannose; Fuc–fucose; Sia–sialic acid; GlcNAc–N-acetylglucosamine; GalNAc–N-acetylgalactosamine; Ser–serine; Thr–threonine, Asn–aspargine).

The *O-linked glycans* (O-glycans) are linked to the hydroxyl group of serine (Ser) or threonine (Thr) side chain. Usually they are short and not significantly branched oligosaccharides, composed of few monosaccharide residues, such as N-acetylgalactosamine, galactose, N-acetylglucosamine, fucose, and N-acetylneuraminic acid, also called sialic acid [23]. The N-linked glycans form the bond with an amide group of the side chain asparagine (Asn). Oppositely to O-linked glycans, *N-linked glycans* are often much larger (longer, more branched) and contain N-acetylglucosamine, mannose, galactose, fucose, N-acetylgalactosamine, and N-acetylneuraminic acid. Their size varies between 6–15 monosaccharide units, arranged in 2 to 5 antennae structures.

The glycans fulfill many important, structural and functional roles. They are responsible for increased solubility and stability of a number of proteins, e.g., as in the case of blood plasma proteins. Oligosaccharides differ from proteins and nucleic acids in a few characteristics: they are usually highly branched and their monomeric units are bound by variety of bonds. Due to the branching and numerous alternative linkages, these types of oligosaccharides are able to carry more information than other biological molecules. Therefore, their varying structural motives serve as a part of a recognition system as in antigen–antibody and receptor–hormone reactions, bacterial infection, cell–cell, and cell–extracellular protein interactions [22]. Unique glycans are also involved in promoting the progression of various carcinomas, for example, some tumor antigens consist of truncated O-linked glycans. Their accumulation in many tumors correlates with invasion. Also, certain glycosphingolipids containing sialic-acids (called *gangliosides*) have been correlated with tumor growth.

2.4 Cytoskeleton

The cell cytoskeleton is a 3D intracellular network of filamentous polymers that provides a continuous intracellular mechanical coupling [1]. It is mainly composed of three main structural elements: (i) microfilaments (MFs), (ii) microtubules (MTs), and

(iii) intermediate filaments (IFs). Major filamentous proteins of the cytoskeleton are actin, tubulin, and intermediate filament proteins. These proteins assemble into linear polymers of variable length: actin forms microfilaments (also called actin filaments), tubulin creates microtubules, and intermediate filament proteins form intermediate filaments. The cytoskeleton is a highly dynamic structure that undergoes constant restructuring, remodeling, and modification in response to environmental and intracellular signals.

All three cytoskeletal components have distinct subcellular localizations. Actin filaments are usually present in a layer known as the "*cell cortex,*" immediately beneath the plasma membrane, and in cell projections such as microvilli. They are also associated with a number of proteins and other structures, such as the contractile ring in dividing cells or focal contacts linking cytoskeleton with surface receptors. Microtubules extend from the nucleus region towards the cell periphery. Intermediate filaments are distributed in a similar pattern to microtubules. IFs and MTs are excluded from the actively expanding leading edge of the moving or "ruffling" cell [1].

2.4.1 Actin Filaments

Microfilaments (μF, [1]) are linear filaments built up from the globular protein actin (G-actin, molecular weight of 43 kDa). These actin molecules polymerize to form a long fiber. Two such fibers are twisted around each other in a helix to form a single actin filament with a diameter of about 7 nm. Within a cell, there are two types of actin filament organization. They form either a mesh composed of short filaments or long, parallel bundles of fibers (Fig. 2.16).

Microfilaments, together with other types of cytoskeletal filaments, are responsible for mechanical strength of the cell. They participate in such processes as cell division into two daughter cells, where their role is to anchor centrosomes at opposite poles of the cell during mitosis or in cell migration. The actin filaments participate also in the linkage of transmembrane (e.g., cell surface receptors) to cytoplasmic proteins.

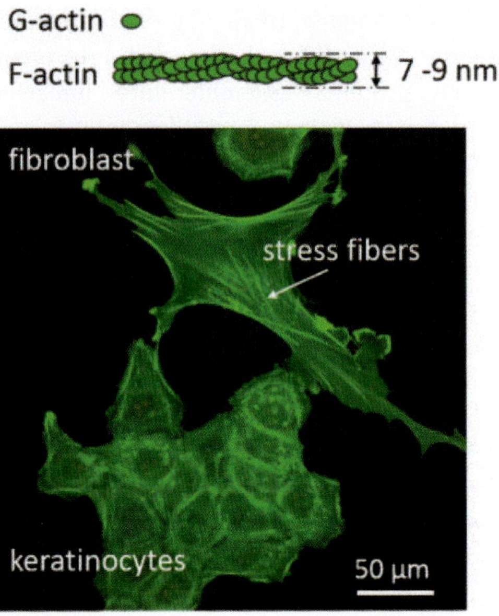

Figure 2.16 The organization of actin filaments in two types of epithelial cells, namely, fibroblasts and keratinocytes, visualized using phalloidin fluorescently labeled using Alexa Fluor 488. Unpublished data courtesy of Barbara Orzechowska, IFJ PAN.

2.4.2 Microtubules

Microtubules (MT, [1]) are small tubes built of tubulin dimmer, composed of two subunits: α- and β-tubulin, both with molecular weight of 55 kDa. Tubulins are organized in a so-called *protofilaments* that further form microtubule (Fig. 2.17). Each microtubule is typically composed of 13 protofilaments arranged around the circumference with diameter of about 24 nm and length in the range of 200 nm to 25 µm. A microtubule extends by the addition of tubulin proteins to one of its ends.

Microtubules are found in the cytoplasm of all eukaryotic cells where they are often observed to spread out radially from a microtubule-organizing center (MTOC) located near the nucleus (Fig. 2.17). The microtubules provide a strong scaffold that supports the cell and determines its shape. Most studies have concluded that microtubules play a positive role by regulating actin polymerization, transporting membrane vesicles or other organelles inside the cell, and/or facilitating the turnover of adhesion plaques. They can also form specialized structures such as centrioles, cilia, and flagella. Both cilia and flagella are cellular appendages, consisting of a core of microtubules enclosed in an extension of the plasma membrane, playing an important role in cellular locomotion.

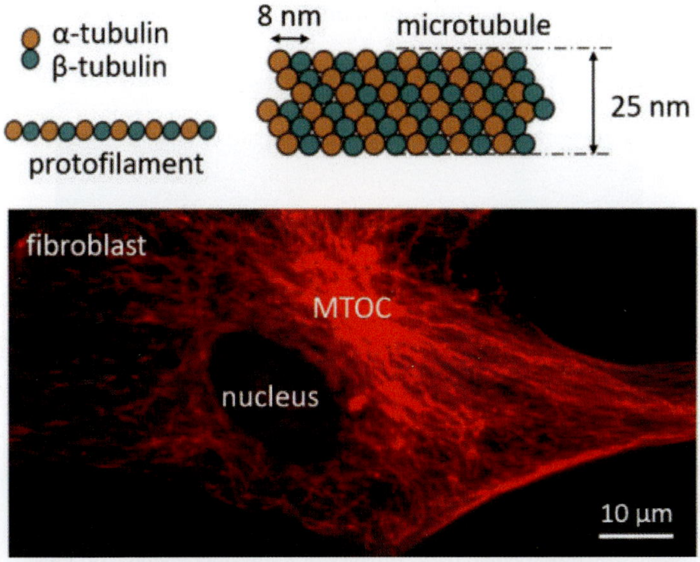

Figure 2.17 The organization of microtubules in a fibroblast. Microtubules were stained with secondary mouse monoclonal antibody labeled with Alexa Fluor 555 (MTOC—the microtubule-organizing center, unpublished data of the author).

2.4.3 Intermediate Filaments

Intermediate filaments (IFs) [1] are the third type of fibrous cytoskeletal components. Their diameter is about 10 nm so they are typically intermediate in size between microfilaments and microtubules. Unlike the microfilaments and microtubules, the intermediate filaments are made of several different proteins. Therefore, the intermediate filaments can be divided into five major types. Type I and II are composed of acidic and basic keratin, respectively. They are produced by different types of epithelial cells (i.e., bladder, skin). Type III encompasses intermediate filaments distributed in a number of cell types, including vimentin in fibroblasts, endothelial cells and leukocytes; desmin in muscle; glial fibrillary acidic factor in astrocytes and other types of glia; and peripherin in peripheral nerve fibers. Type IV are neurofilaments and type V are made of laminin (Fig. 2.18).

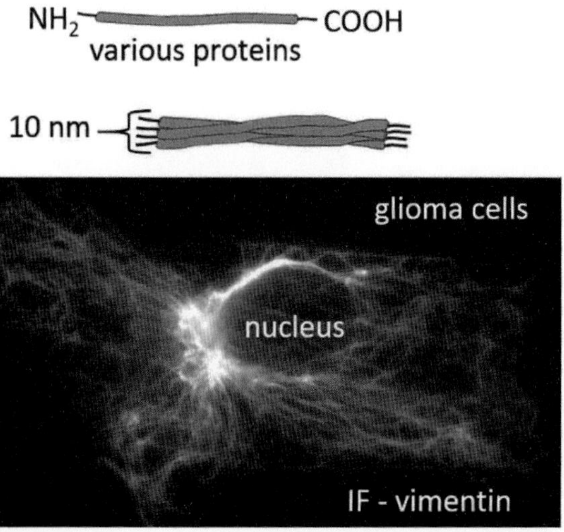

Figure 2.18 Distribution of intermediate filaments (composed of vimentin) in primary astrocytes. Scale bar: 10 μm. Unpublished data courtesy of Katarzyna Pogoda, IFJ PAN.

Each microtubule is typically composed of 13 protofilaments arranged around the circumference with diameter of about 24 nm and length in the range of 200 nm to 25 μm. A microtubule extends by the addition of tubulin proteins to one of its ends. Microtubules are found in the cytoplasm of all eukaryotic cells where they are often observed to spread out radially from a centrosome located near the nucleus. From this center, they provide a strong frame that supports the cell and determines its shape. They also serve as tracks along which cellular organelles can migrate. Microtubules can also form specialized structures such as centrioles, cilia, and flagella. Both cilia and flagella are cellular appendages consisting of a core of microtubules enclosed in an extension of the plasma membrane, playing an important role in cellular locomotion.

References

1. Lodish, H., Berk, A., Matsudaira, P., Kaiser, C. A., Krieger, M., Scott, M. P., Zipursky, S. L., and Darnell, J. (2004). *Molecular Cell Biology*, 5th ed. (W. H. Freeman and Company, New York, USA).
2. Nelson D. L., and Cox, M. M. (2003). *Lehringer Principles of Biochemistry*, 3rd ed. (Worth Publishers, New York, USA).
3. Bateman, J. F., Boot-Handford, R. P., and Lamande, S. R. (2009). Genetic diseases of connective tissues: Cellular and extracellular effects of ECM mutations. *Nat. Rev. Genet.*, **10**, pp. 173–183.
4. Hynes, R. O. (2009). The extracellular matrix: Not just pretty fibrils. *Science*, **326**, pp. 1216–1219.
5. Hynes, R. O., and Naba, A. (2012). Overview of the matrisome–an inventory of extracellular matrix constituents and functions. *Cold Spring Harb. Perspect. Biol.*, **4**, p. a004903.
6. Buehler, M. J. (2006). Nature designs tough collagen: Explaining the nanostructure of collagen fibrils. **103**, pp. 12285–12290.
7. Weaver, S. A., Wolters, B., Ito, N., Woskowicz, A. M., Kaneko, K., Shitomi, Y., Seiki, M., and Itoh, Y. (2014). Basal localization of MT1-MMP is essential for epithelial cell morphogenesis in 3D collagen matrix. *J. Cell. Sci.*, **127**, pp. 1203–1213.
8. Potts, J. R., and Campbell, I. D. (1994). Fibronectin structure and assembly. *Curr. Opin. Cell Biol.*, **6**, pp. 648–655.
9. Krammer, A., Lu, H., Isralewitz, B., Schulten, K., and Vogel, V. (1999). Forced unfolding of the fibronectin type III module reveals a tensile molecular recognition switch. *Proc. Natl. Acad. Sci. U. S. A.*, **96**, pp. 1351–1356.
10. Santos-Valle, P., Guijrro-Munoz, I., Cuesta, A. M., Alonso-Camino, V., Villate, M., Alvarez-Cienfuegos, A., Blanco, F. J., Sanz, L., and Alvarez-Vallina, L. (2012). Heterotrimeric laminin coiled-coil domain exerts anti-adhesive effects and induces a pro-invasive phenotype. *PLoS One*, **7**(6), p. e39097.
11. Sarrazin, S., Lamanna, W. C., and Esko, J. D. (2011). Heparan sulfate proteoglycans. *Cold Spring Harb. Perspect. Biol.*, **3**(7), p. a004952.
12. Responte, D. J., Natoli, R. M., and Athanasiou, K. A. (2012). Identification of potential biophysical and molecular signaling mechanisms underlying hyaluronic acid enhancement of cartilage formation. *J. R. Soc. Interface*, **9**, pp. 3564–3573.

13. Van Meer, G., Voelker, D. R., and Feigenson, G. W. (2008). Membrane lipids: Where they are and how they behave. *Nat. Rev. Mol. Cell Biol.*, **9**, pp. 112–124.

14. Plow, E. F., Haas, T. A., Zhang, L., Loftus, J., and Smith, J. W. (2000). Ligand binding to integrins. *J. Biol. Chem.*, **275**, pp. 21785–21788.

15. Leckband D. E., and Prakasam, A. (2006). Mechanism and Dynamics of Cadherin Adhesion. *Annu. Rev. Biomed. Eng.*, **8**, pp. 259–287.

16. Fouquet, B., Zimbelmann, R., and Franke, W. W. (1992). Identification of plakoglobin in oocytes and early embryos of Xenopus laevis: Maternal expression of a gene encoding a junctional plaque protein. *Differentiation*, **51**, pp. 187–194.

17. Ramburan, A., and Govender, D. (2002). Cadherins and catenins in pathology. *Curr. Diagn. Pathol.*, **8**, pp. 305–317.

18. Nollet, F., Berx, G., and vanRoy, F. (1999). The role of the E-cadherin/catenin adhesion complex in the development and progression of cancer. *Mol. Cell Biol. Res. Commun.*, **2**, pp. 77–85.

19. Brasch, J., Harrison, O. J., Honig, B., and Shapiro, L. (2012). Thinking outside the cell: how cadherins drive adhesion. *Trends Cell Biol.*, **22**, pp. 229–310.

20. Varki, A. (1994). Selectin ligands. *Proc. Natl. Acad. Sci.*, **91**, pp. 7390–7397.

21. Evans, E. (1995). Physical actions in biological adhesion, in: *Structure and Dynamics of Membranes*, **1**, (Lipowsky, R., and Sackmann, E., eds.), Chapter 15, Elsevier Science B. V. Amsterdam, pp. 723–754.

22. Goochee C. F., Gramer, M. J., Andersen, D. C., Bahr, J. B., and Rasmussen, J. H. (1991). The oligosaccharides of glycoproteins: Bioprocess factors affecting oligosaccharide structure and their effect on glycoprotein properties. *Biotechnology*, **9**, pp. 1347–1355.

23. Mechref, Y., and Novotny, N. V. (2002). Structural investigations of glycoconjugates at high sensitivity. *Chem. Rev.*, **102**, pp. 321–369.

Chapter 3

Principles of Atomic Force Microscopy

There are several techniques that allow to study the properties of living cells (i.e., their ability to deform and to adhere) at the single cell and molecular levels. Among them, the noteworthy ones are those delivering information of local character, such as atomic force microscopy (AFM, [1]). In recent years, an ample evidence has demonstrated the functionality of this technique to characterize changes of single, living cells—based on either their elastic properties or single molecule interactions in respect of alterations in cellular functionality, structure, etc. This chapter introduces basic operational aspects of atomic force microscopy.

3.1 Principles of the AFM Operation

The idea of the AFM operation is very straightforward, but its realization requires highly advanced technology. A sharp, delicate probing tip, mounted at the end of a compliant cantilever, is moved in close proximity over an investigated surface. The probing tip senses forces acting within the contact area. These forces cause the deflection of a compliant cantilever that is monitored and recorded by a detector system. Independently of the technical details of AFM construction, there are three basic

Cellular Analysis by Atomic Force Microscopy
Malgorzata Lekka

ISBN 978-981-4669-67-2 (Hardcover), 978-1-315-36480-3 (eBook)
www.panstanford.com

elements common for all types of apparatus: (i) a cantilever with a probing tip, (ii) a system detecting the cantilever deflection, and (iii) a system that provides scanning and positioning.

Figure 3.1 presents a scheme of the AFM working in liquid conditions, where the use of a so-called *"liquid cell"* is strongly needed. The scheme shows also the most common cantilever detection system, composed of a laser and a position-sensitive photodetector (i.e., photodiode). Initially, a laser beam is focused at the free end of a cantilever. Then, in a close proximity to the investigated surface, interaction forces cause the cantilever deflection, thus moving the spot position of the reflected laser beam within an active area of the photodiode. The position of the laser beam spot, delivering the information on the cantilever displacement z, can be converted into force F using Hooke's law:

$$F = k_{\text{cant}} \cdot z \tag{3.1}$$

where k_{cant} denotes the cantilever spring constant, expressed in N/m.

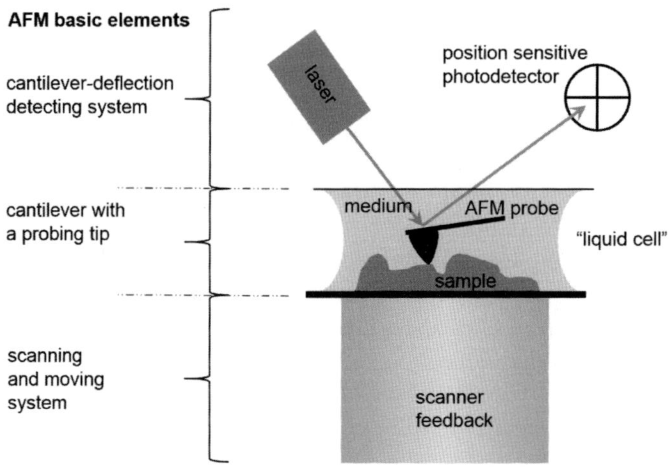

Figure 3.1 Basic elements of the atomic force microscope using optical detection system. Independently of the AFM device type, there are three basic elements, namely: a cantilever with a probing tip, and two systems—one that detects cantilever deflection, and the other one that provides scanning and positioning (since the principle of operation is not dependent on the environment, i.e., ambient conditions or liquid, a "liquid cell" setup is schematically shown).

The sample, attached strongly to a support (like mica or glass coverslip), is mounted on a holder fixed to a piezoelectric scanner. The scanner provides the possibility to perform a raster scan over the investigated sample surface. Usually, the piezoelectric scanner is mounted on top of the coarse positioning system facilitating a convenient and rapid sample exchange.

3.1.1 Cantilevers

Cantilevers are very often made of silicon or silicon nitride. The basic lever with a tip is mounted on a rectangular chip, which enables easy handling (Fig. 3.2).

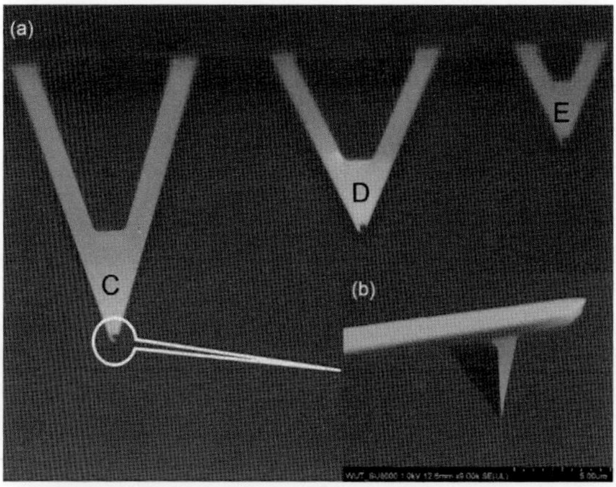

Figure 3.2 (a) An image of a silicon nitride chip with several cantilevers characterized by various spring constants (MLCT cantilevers type C, D and E with the corresponding spring constant of 0.01, 0.03 and 0.1 N/m). Inset: (b) The electron microscope image of a probing tip at the end of the cantilever. Image of the cantilever tip—courtesy of Piotr Bobrowski.

Each single cantilever is characterized by its material properties and geometry: length, width, thickness, and shape (it can be rectangular or triangular one). These parameters define cantilevers' elastic properties that are quantitatively described by a spring constant expressed in N/m. Typical geometries of the silicon nitride cantilevers are shown in Table 3.1. At the end of the cantilever, a probing tip is located. Very frequently, the probing tip possesses the pyramid-like shape (Fig. 3.2b), however, it can also be in a form of a cone or a sphere. Independently of the tip shape, the radius of curvature is defined as the radius of a circular sphere that best approximates the tip end. Typically it ranges from 2 to 50 nm.

Table 3.1 Geometrical parameters of exemplary silicon nitride cantilevers (MLCT type), frequently used in the measurements of biological samples

Probe Type—MLCT						
Cantilever type	**A**	**B**	**C**	**D**	**E**	**F**
Nominal spring constant [N/m]	0.07	0.02	0.01	0.03	0.1	0.6
Resonant frequency [kHz]	22	15	7	15	38	125
Length [μm]	175	210	310	225	140	85
Width [μm]	22	20	20	20	18	180

Note: The nominal cantilever thickness is 0.55 μm; however, according to manufacturer's data, thickness varies from 0.5 to 0.6 μm.

It has always been clear that the tip shape of the AFM probe influences the recorded images. Thus, acquired images are a convolution of the properties of the sample and the AFM probe. This is illustrated in Fig. 3.3, showing an idealized contact-mode AFM experiment carried out in two extreme conditions.

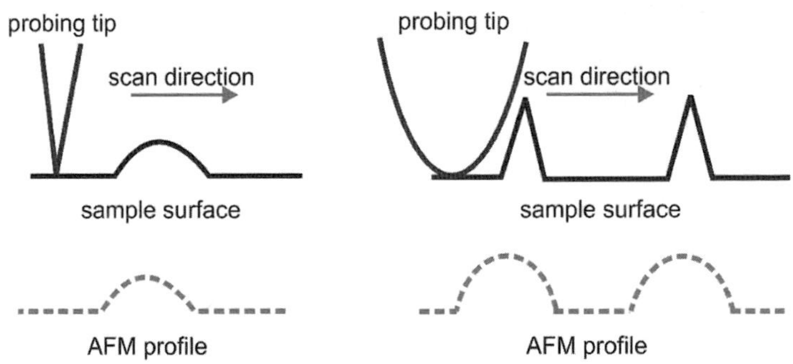

Figure 3.3 The influence of the AFM probing tip geometry on recorded images. If the radius of curvature is much smaller than the size of the studied surface structures, the measured AFM profile (red dash lines) follows the real shape of the studied structure. In opposite conditions, the resulting image follows the shape of the probing tip.

If the radius of curvature is much smaller than the structures' size on an investigated surface, the AFM probe traces the real shape of the studied structures (red dash line in Fig. 3.3, recorded by the AFM).

If the radius of curvature is much larger than the surface structures, the AFM profile reveals the shape of the probing tip. In reality, the AFM image reflects both the shape of the surface structures and the shape of the probe. The use of the sharpest possible probes can minimize erroneous features generated due to such a convolution but, simultaneously, can induce surface damages if the sample is delicate.

The effect of the AFM probe size on the imaged surface structures can be easily illustrated during the imaging single protein molecules like concanavalin A (Con A, *unpublished data of the author*, Fig. 3.4a).

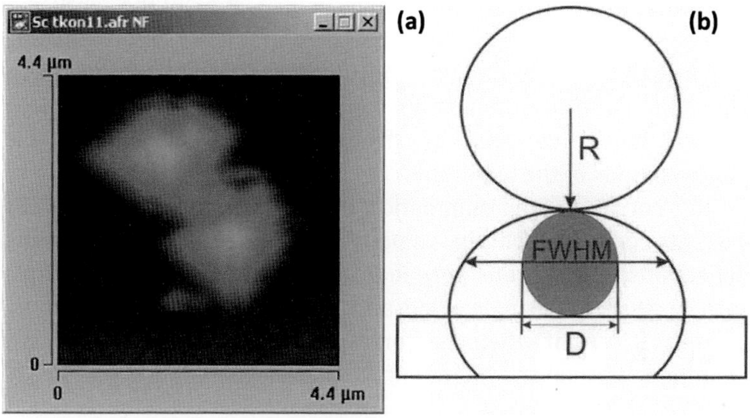

Figure 3.4 (a) The effect of the AFM probe (a four-sided pyramid with open angle of 35° and radius of curvature of 50 nm) on single protein molecule (Con A, concanavalin A from *Canavalia ensiformis*), attached to mica surface and recorded in phosphate buffered saline (pH = 7.4) using a home build AFM system working at the IFJ PAN, Kraków, Poland). (b) The idea of the molecule diameter de-convolution.

Concanavalin A is a glycoprotein isolated from *Canavalia ensiformis*, which was dissolved in phosphate buffered saline (PBS buffer, pH = 7.4). At pH > 7.0, Con A is a globular/ellipsoidal tetramer with a single molecule dimensions of 6.7 nm ×

11.3 nm × 12.2 nm [2]. Clearly visible rectangular shape of Con A indicates the effect of tip shape on the recorded images of molecule topography. In case of globular structures, the deconvolution of the tip shape, and thereby, the determination of real dimensions can be obtained from the following equation [3] (Fig. 3.4b):

$$\mathrm{FWHM} = 2\cdot\sqrt{R\cdot D+\frac{D^2}{4}}, \tag{3.2}$$

where FWHM is the full width taken at half maximum, *R* is the radius of the curvature of the AFM probe, and *D* is the diameter of the studied molecule.

3.1.2 Detection System of Cantilever Deflection

The most common system detecting the deflection of the cantilever is composed of laser and position-sensitive photodetector, which active area is divided into four quadrants (see Inset 1 in Fig. 3.5). Cantilever defection causes the laser beam to move within the active area of the detector, thus changing the photocurrents of the quadrants.

By corresponding summation and subtraction of photodiode currents one can obtain values that are proportional to the deflections in directions perpendicular and parallel to the sample surface. If I_{xy} is a single quadrant current (U = up, B = bottom, L = left, R = right) then the corresponding angle Θ of cantilever bending in normal and lateral directions are

$$\Theta_\mathrm{N} = A\cdot\left[(I_\mathrm{UL}+I_\mathrm{UR})-(I_\mathrm{BL}+I_\mathrm{BR})\right] \tag{3.3}$$

and

$$\Theta_\mathrm{L} = B\cdot\left[(I_\mathrm{UL}+I_\mathrm{BL})-(I_\mathrm{UR}+I_\mathrm{BR})\right], \tag{3.4}$$

where N denotes the normal bending, L is the cantilever twist, and *A* and *B* are proportional coefficients. Very often, in order to minimize the influence of the laser power fluctuations, these signals are normalized by dividing them by a sum of photocurrents ($I_\mathrm{TOT} = I_\mathrm{UL} + I_\mathrm{BL} + I_\mathrm{UR} + I_\mathrm{BR}$).

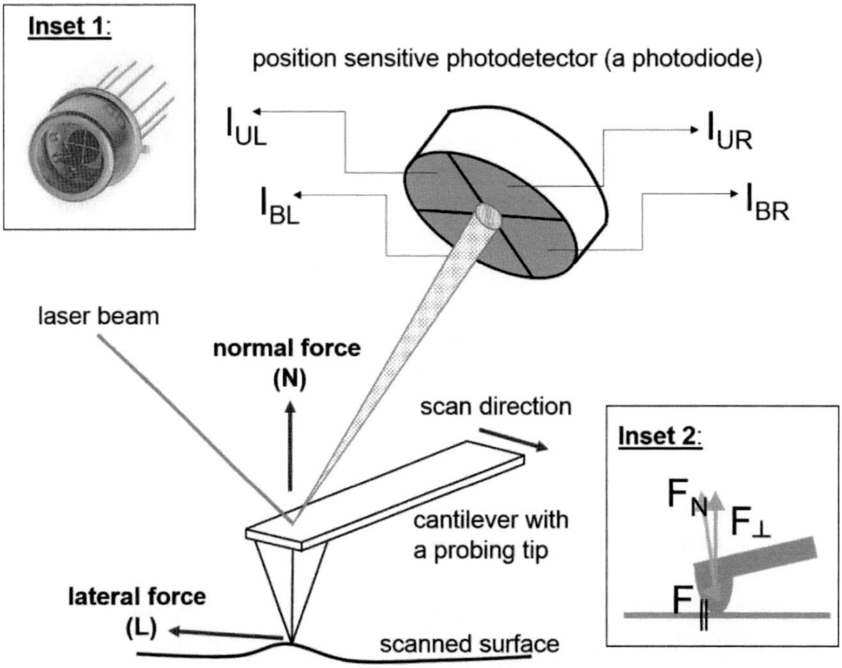

Figure 3.5 The idea of the measurements of normal and lateral deflections of the cantilever. Inset 1: a photograph of a position-sensitive detector (i.e., a photodiode). Inset 2: components of the force acting on the cantilever.

The cantilever bending linearly approximates the deflections both in normal and lateral directions. Knowing mechanical properties of the cantilever, it is possible to estimate the values of normal and lateral forces (F_N and F_L, respectively). Taking into account the geometrical shape of cantilever, much higher sensitivity of the normal forces is expected, i.e., cantilever bends much easier in normal direction. One has also note that it is impossible to place the cantilever exactly parallel to the investigated surface. Because of technical reasons, cantilevers are usually tilted by about 10–12° to the surface. This denotes that the overall force deflecting the cantilever in normal direction will, i.e., perpendicular and parallel to the surface (see Inset 2 in Fig. 3.5).

3.1.3 Feedback Loop

Commercially available AFM systems are equipped with a *feedback loop* that is used to control and maintain a constant interaction between the probing tip and sample surface during its scanning. The feedback loop is applied to monitor cantilever deflection and then to adjust the height of the cantilever, thus taking into account the changes in sample surface height. Following the feedback signal, the base of the cantilever is moved up and down over higher and lower parts of the sample to provide the same force value. In such way, if the system is properly set up, the maximum force applied to sample can be controlled.

Practically, the AFM system monitors vertical deflection. The user specifies a desired value to be maintained (called *a setpoint*). The difference between the setpoint and the actual deflection is measured and referred to as *the error*. This error is scaled by the gain, and this signal is used to control the movement of the scanner. If the gain is too low, the system will not respond fast enough to changes in sample topography. As a result, the image quality will be poor and the tip will get overexploit. If the gain is set to high, the feedback loop will become unstable and the recorded image will appear to be excessively noisy. Only if the error signal in the feedback loop is properly scaled, the system responds sufficiently quickly to changes in topography, thus keeping the error signal minimized.

To control the feedback loop, two parameters are applied (1) a time constant (called *integral gain*) and (2) a value proportional to the magnitude of the error signal (called *proportional gain*). These two parameters regulate how quickly the feedback responds to a change in the sample height. In practice, the integral gain is usually raised until the image becomes noisy, and then the gain is reduced slightly. However, both parameters have to be optimized for different imaging conditions, depending, for instance, on the sample topography and scanning speed.

3.1.4 Scanning and Positioning System

In the AFM, the precise manipulation of the sample and its scanning is executed using piezoelectric devices that provide adequate displacement sensitivity (from few to tenths of nanometers per volts) and fast initial response times (~10 μs). The piezoelectricity is a phenomenon in which the electric charge is accumulated in certain materials in response to mechanical stress. This direct piezoelectric effect converts mechanical energy into the electrical one. The effect is reversible—the applied voltage causes a change in length (this is called *inverse piezoelectric effect*), thus electrical energy is converted into the mechanical one. Careful control of the voltage applied to piezoelectric material, provides small displacements of the material.

Early constructions of the scanning system were based on a tripod (separate transducers fixed in orthogonal directions) or a bimorph (two piezoelectric layers bonded to a thin metal shim sandwiched in the middle) arrangements. Most commonly, scanners have a form of a tube made of lead zirconate titanate (Fig. 3.6).

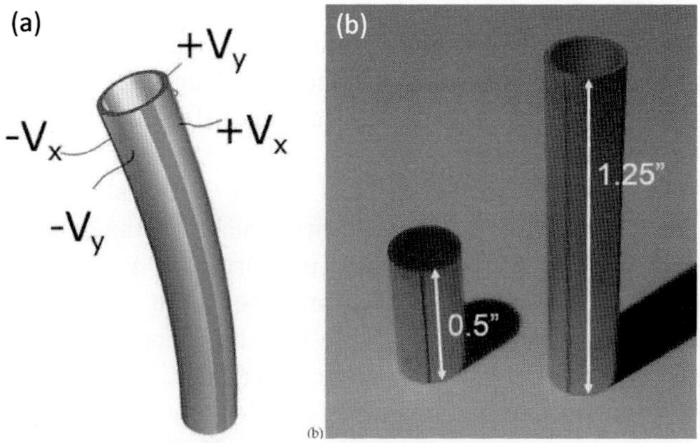

Figure 3.6 (a) Scheme of the piezoelectric scanner (tube) with outer surface divided to four electrodes to which positive and negative voltages *V* are applied. (b) A photo of two typical piezo scanners.

The scanner (termed also *a piezo scanner*) has the outer surface segmented parallel to the tube axis into four electrodes (Fig. 3.6a). The tube interior wall serves as an internal electrode. A voltage applied to each electrode causes an appropriate bending depending on the magnitude and sign of the given voltage.

$$\Delta L = \pm \cdot d_{ij} \cdot L_0, \tag{3.5}$$

where ΔL is the change in length [m], L_0 is the initial length of the piezoelectric tube [m], and d_{ij} is the piezoelectric coefficient of the material [m/V]. The extension or contraction in Z-direction is accomplished by applying positive or negative voltage to all four electrodes. This leads to changes in piezo scanner length L and radius R:

$$\Delta L = \frac{d_{31} \cdot V \cdot L_0}{w} \tag{3.6}$$

and

$$\Delta R = d_{33} \cdot V, \tag{3.7}$$

where V is the applied voltage, w is the wall thickness of the tube, and d_{31} and d_{33} are directional (tensor) coefficients characterizing the piezo material: d_{33} describes the strain parallel to the polarization vector of the ceramics (thickness) used when calculating the displacement of stack actuators; d_{31} is the strain orthogonal to the polarization vector (width) used for calculating tube and strip actuators.

After applying voltage to one of the outer electrodes, the tube bends in the direction perpendicular to the length axis. The magnitude of bending is proportional to the voltage difference between the common and the outer electrode. Thus, precise movement in the XY plane may be executed by applying voltage to two neighboring outer electrodes. Additionally, the scanning range may be enhanced (doubled) by applying voltages of opposite sign to the opposite electrodes. Maximum scan range S realized by a single quadrant under the applied voltage V may be calculated using the following formulae [4]:

$$S = \frac{\sqrt{2} d_{31} V L^2}{\pi R w} \tag{3.8}$$

The XY motion is not entirely orthogonal to Z-movement. Scanning is achieved by piezoelectric scanner bending, therefore, a scanner of the form of a tube scans the surface of a sphere rather than that of a plane [5]. The estimation for the Z deflection (referred as a *distortion*) from the XY plane is as follows:

$$\Delta h = 0.5 \cdot \frac{\Delta S^2}{L} \tag{3.9}$$

The magnitude of the distortion is dependent on the piezoelectric tube length and the size of the scan area. For example, L = 1 cm and scanning range S = 1 μm, an error in the Z-direction is only 0.05 nm. Some of the currently available AFM systems do not use piezoelectric tube to move sample in XYZ directions, rather the XY movement is separated from the Z one by additional piezoelectric scanner that decouples Z motion from the X and Y scanning. This enables to eliminate a surface curvature for a wide variety of sample types and sizes, and it provides a flat, highly linear and orthogonal XY scan, with an accurate and precise angle measurement.

Depending on the sign of the applied voltage, both extension and contraction can be realized. The idea of how the tube moves in the AFM is presented in Fig. 3.7.

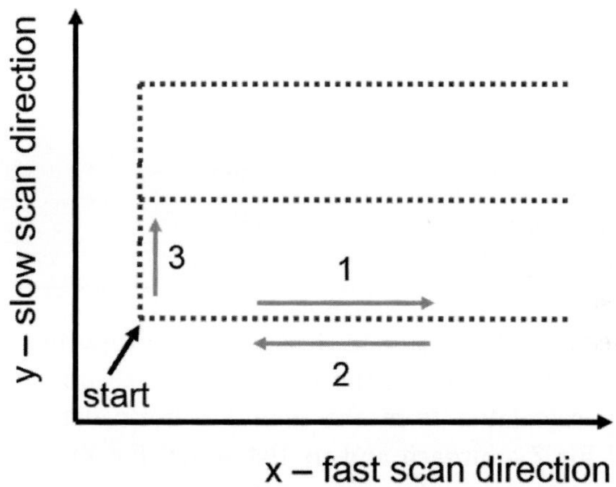

Figure 3.7 The idea of a scan realized in the AFM. Starting from zero position, the voltage is applied leading to tube bending along x-axis (fast scan direction (1)), followed by voltage withdrawing, scanner is returning to zero position on the same way (2). After coming back, the voltage is updated in y-axis (slow scan) direction (3). The voltage step is related to the scan resolution.

Beginning with the "start" position (maximum bending both in X and Y directions), the voltage of scanner electrodes responsible for tube bending along *X*-axis (called *fast scan direction*) is increased up to a maximum value. When voltage is removed, the tube returns to the "start" position along the same path. Next, voltage of electrodes responsible for the bending along *Y*-axis (*slow scan direction*) is updated, followed by the applying of voltage in fast scan direction, and so on.

The properties of an exemplary piezoscanner made of lead zirconate titanate are presented in Table 3.2.

Table 3.2 Properties of an exemplary piezoelectric scanner made of lead zirconate titanate (the piezoelectric scanner type EBL #2 from Stavely Sensors, USA)

Material	Lead zirconate titanate
Geometry	
Length	0.5 or 1 inch
Diameter	0.25 inch
Wall thickness	0.02 inch
Properties	
Piezoelectric coefficient d_{31}	-173×10^{-12} m/V
Piezoelectric coefficient d_{33}	380×10^{-12} m/V
Max. Voltage	300 V
Curie temperature	350°
Frequency	53 kHz

For piezoelectric scanner, described by properties presented in Table 3.2, in the case of the length of the 0.5 inch, the voltage sensitivity calculated from the manufacturer's data is equal to 4.3 nm/V for *Z* approach and to 15.6 nm/V for *XY* scanning. For example, by applying the voltage of 100 V, one can achieve the extension ΔL = 0.43 μm and the full scan range $2S$ = 3.1 μm. High frequency constant ensures good mechanical properties, while high Curie temperature allows applications in the UHV environment, where baking of the whole system is required.

Piezoelectric scanners are far from ideal and suffer from many parasitic effects [6]. Main sources of distortion are nonlinear voltage response and hysteresis (Fig. 3.8). Hysteresis and scanner nonlinearity cause various artifacts in scanned images that can be easily observed in topography images of a calibration grating.

Nonlinear response introduces size errors, especially at higher voltages needed for large-scale scans (~1 μm), while in the case of small features in AFM imaging the assumption of linear behavior can be justified. This effect may be minimized when the function describing the response of the piezoelectric scanner is known. Then, the voltage value corrections may be applied online during the scan.

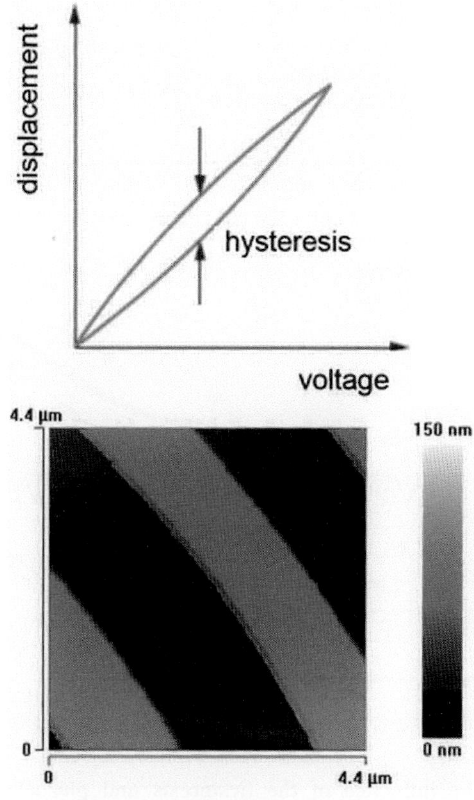

Figure 3.8 The hysteresis and non-linearity of piezoelectric scanner is observed when the piezoelectric scanner response (e.g., its displacement) induced by the applied voltage has non-linear character. Actually, the imaged steps of a calibration grating (right image) are linear and parallel.

The influence of the piezoelectric scanner hysteresis and non-linearity is also observed in measurements carried out in the force spectroscopy mode.

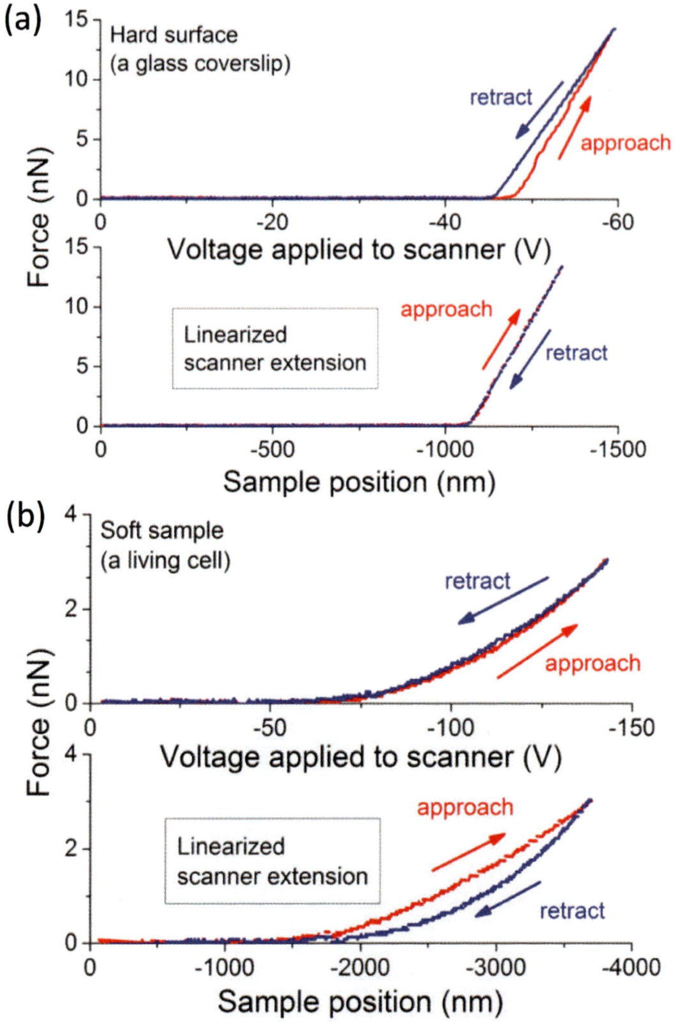

Figure 3.9 The influence of the hysteresis and piezoelectric scanner nonlinearity in force curves recorded using AFM force spectroscopy mode on (a) a hard surface (a glass coverslip) and (b) on soft sample (a living cell). Reprinted with permission from [9].

The force curves recorded with a non-linearized scanner show distinct behavior depending on their mechanical properties. In case of stiff, non-deformable surfaces, when a non-linearized scanner is used, the approach part of the force curve does not overlap with the retract one (Fig. 3.9a). After the correction, both curves should overlap. Different behavior can be observed for soft samples, as living cells. Here, the scanner nonlinearity may lead to totally wrong interpretation. In the upper plot of Fig. 3.9b, both parts of the force curve overlap, which suggests a pure elastic interaction of the AFM probe with a living cells. However, after the correction, true data shows the presence of an approach–retract hysteresis, resulting from viscoelastic character of the studied sample.

3.2 Force Spectroscopy

To investigate the sample properties by the AFM, the force spectroscopy mode is widely applied [7, 8]. In this mode, information about the sample properties is derived from deflection changes as a function of a cantilever separation from the surface, recorded at a single point rather than by scanning the lateral position of the tip (from a so-called *force curve*). In the single force curve, the base of the cantilever is moved in the vertical direction towards the surface and then retracted again. During the motion, the deflection of the cantilever is recorded as a function of relative sample position (Fig. 3.10).

The force curve consists of two parts: the one recorded during approaching the tip to the sample surface (an approach curve, red line in Fig. 3.10) and the other one, collected during opposite motion (a retract curve, black line in Fig. 3.10). The character (shape) of the force curve depends on physical and chemical properties of two interacting surfaces, namely the investigated sample, the probing tip, and the surrounding environment.

Regardless of the sample type, there are several features that are present in all curves. When the cantilever is away from the surface, its deflection should be zero since there is no detectable interaction force. Actually, due to thermal vibrations, the cantilever oscillates around its free position, reflecting the noise present in a particular AFM system (this is represented by a horizontal *base line* (A)). During the approach, if both the tip and the surface are charged with the same sign, at close distances—prior to the contact—the cantilever can be repulsed from the surface (*region R*). It is represented by the slight raise in the baseline, as visible in

Fig. 3.10. The presence of attractive forces between the tip and a surface is reflected by a jump-in (*point B*), i.e., the moment when the cantilever is suddenly attracted. At this moment, the gradient of the attractive force is larger than the cantilever spring constant. When the tip is in contact with the sample surface, the electron clouds of atoms of both the tip and the sample are overlapping and repulsing. The further approach results in the cantilever bending (up to the certain maximum value, *point C*), which character depends on the material properties of the investigated sample (linear or not). During the retraction, the interacting repulsive force decreases and, the tip does not separate from the surface exactly at the same point where it started to touch the surface. Forces that are responsible for such behavior arise from adhesive properties of investigated surfaces. In further retract when the elastic force of the cantilever exceeds the gradient of the adhesive force, the tip is rapidly separated from the surface. The *point D* corresponds to the maximum value of the force (a so-called *pull-off force*). Further separation yields in cantilever fluctuations around its free position (base line).

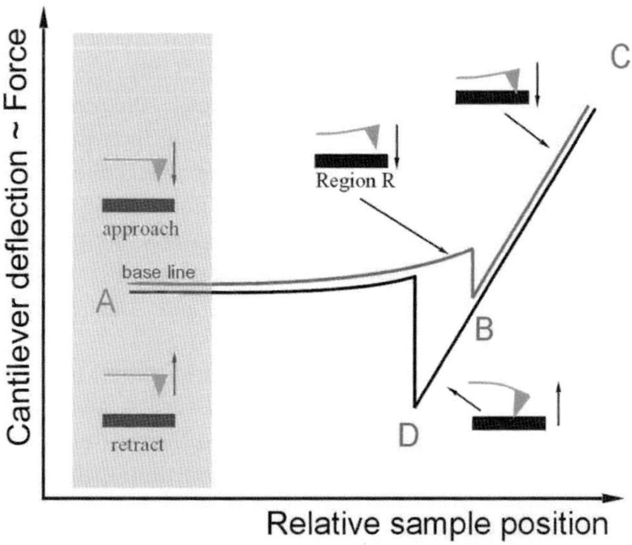

Figure 3.10 A schematic view of the force curve recorded by the AFM working in the force spectroscopy mode. Arrows indicate direction of the movement of the AFM cantilever. Adapted with permission from [9].

3.2.1 Calibration

In most AFM systems, the deflection of the cantilever is measured as a difference between voltages obtained from position-sensitive detector system. To convert the deflection into distance or force units (*newtons*), the recorded signal (*volts*) is multiplied by two factors: a photo detector sensitivity (*PSD*) and cantilever spring constant (k_{cant}). The former converts *volts* into *nanometers,* while the latter enables to determine the force value in *nanonewtons* according to

$$F[\text{nN}] = \text{Signal}[\text{V}] \cdot \text{invPSD}\left[\frac{\text{nm}}{\text{V}}\right] \cdot k_{\text{cant}}\left[\frac{\text{N}}{\text{m}}\right] \tag{3.10}$$

In reality, to determine the force, the inverse of the PSD (invPSD) is widely applied, since it directly converts nanometers to volts. The proper calibration of the AFM system is crucial for collection of reliable quantitative data. Therefore, both PSD and spring constant should be determined prior to use of the AFM.

3.2.1.1 Photodetector sensitivity (PSD)

The deflection of the cantilever is recorded using the optical system composed of a laser and a position-sensitive detector. Thus, the deflection is registered as a difference between quadrant voltages and it is expressed in volts. Typically, to calibrate the cantilever deflection, the force curve is recorded at a stiff, non-deformable surface. For such surfaces, after reaching the contact, the deflection directly reflects the position of the sample (represented by a linear sloped curve, Fig. 3.11).

Usually, due to surface repulsion and/or contamination, the fraction (around 10–20%) of the linear part of the approach just after contact is excluded from the linear regression. In many biological applications, where the force spectroscopy mode is involved, the commonly applied cantilevers have the spring constant within the range of 0.01–0.1 N/m and the typical loading force value does not exceed 30 nN [10]. Assuming that the typical silicon nitride cantilever with a paraboloidal shape (tip radius of 20 nm) indents the flat surface with the load force of 30 nN, the resulting indentation depth is merely of 0.2 nm for glass (this estimation is performed assuming the Hertz mechanics of contact) [10–12]. Therefore, in this range, glass surface can be used as a stiff, non-deformable sample.

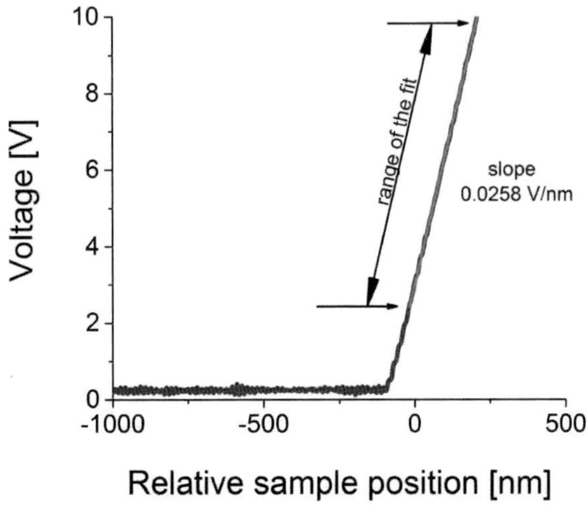

Figure 3.11 The approach part of the force curve, recorded on a glass coverslip (a stiff, non-deformable surface). The *photodetector sensitivity* (*PSD*, a conversion factor from volts to nanometers), was determined as a slope of the fitted line (reliable fit range marked by arrows).

The invPSD coefficient is determined as an inverse value of the slope obtained from the linear fit to the region where, after the contact with the surface, the deflection directly reflects the position of the sample. The obtained value reflects the properties of the optical system used for the cantilever deflection. However, one should keep in mind that it changes not only from one AFM system to the other but also from one measurement to another. Fortunately, for a particular AFM system, careful cantilever treatment (especially, during liquid exchange), and proper settings of the optical system assures that its value does not change much (Fig. 3.12).

The invPSD values, determined from a Gaussian fit to obtained distributions for the specific cantilever type (MLCT and OTR-4, with nominal spring constants of 0.01 and of 0.02 N/m, respectively), corresponds to 38.5 ± 2.7 and to 77.3 ± 5.7 nm/V. The half width of the distribution at half-maximum height (HWHM), used as an error estimate, is below 10% of the central value for 0.01 N/m and 0.02 N/m cantilevers. Much lower and much larger values of the invPSD conversion factor may indicate broken cantilever. Therefore, in such case, the cantilever must be exchanged.

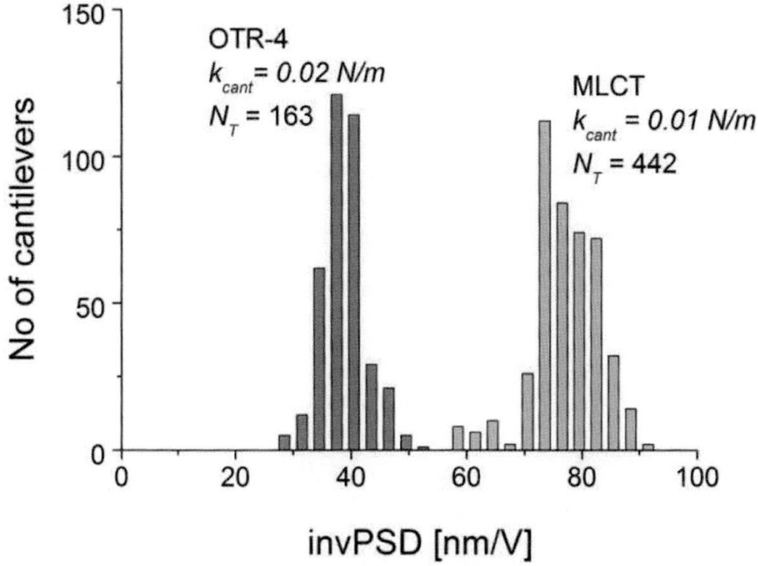

Figure 3.12 The distribution of the invPSD values, determined for two cantilever types, MLCT and OTR-4, used in two AFM microscopes (home build one and the Xe120 from Park Systems, respectively). N_T is the total number of cantilevers analyzed. Unpublished data of the author (OTR-4 compared with that from [9] with permission).

3.2.1.2 Correction factor κ for PSD sensitivity

Since cantilevers are not single harmonic oscillators, they exhibit several oscillation modes that are usually distinguishable from their resonant frequency. Moreover, oscillation modes are different for free and fixed (i.e., in contact with the surface) cantilevers, which leads to two distinct values of PSD sensitivity. Since the optical lever method of cantilever deflection detection monitors the deflection angle of the cantilever tip the relationship between the tip angle and the tip displacement must be considered in the interpretation of AFM data. Therefore, a correction factor κ can be introduced as [13, 14]:

$$\kappa = \frac{\text{PSD\_sensitivity}_{\text{free\_cantilever}}}{\text{PSD\_sensitivity}_{\text{fixed\_cantilever}}} \tag{3.11}$$

The resulting correction factor was determined to be of 1.09 [14, 15], under the assumption of an infinitively small laser spot positioned at the very end of a rectangular cantilever.

3.2.1.3 Spring constant

The quantitative force measurements, carried out using force spectroscopy mode, require an accurate knowledge of the cantilever spring constant. Below, the most common methods for the cantilever calibration are presented.

Theoretical calculation based on cantilever geometry

Theoretical calculations can be applied when the geometry and materials constants are given (Fig. 3.13). It is relatively easy to do this if the used cantilever has a rectangular shape [7].

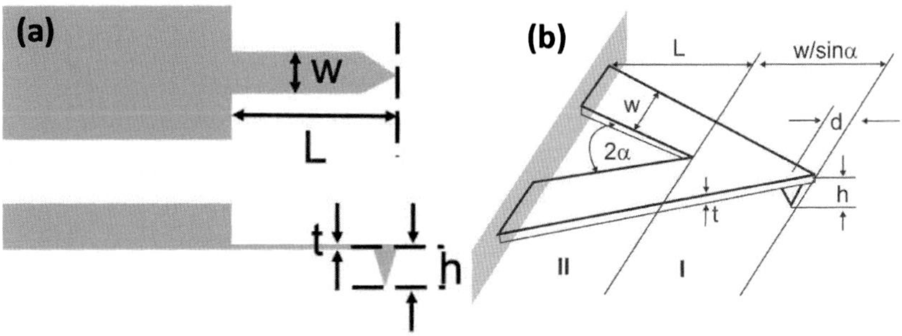

Figure 3.13 The geometry of two common AFM cantilevers geometries: (a) rectangular and (b) triangular with marked triangular plate and two prismatic beams. *L*—length, *w*—width, *t*—thickness; *h*—tip height; *d*—position of the tip center. Adapted with permission from [9].

In case of rectangular cantilever (Fig. 3.13a), the spring constant can be estimated following the equation:

$$k_{\text{cant}} = S\frac{E \cdot w \cdot t^3}{4 \cdot L^3}, \tag{3.12}$$

where *E* is Young's modulus of the material from which cantilever is made of, *L*, *w*, and *t* denote the lever length, width, and thickness, respectively. It is important to remember that the lever thickness present in the numerator is raised to the power of 3. As a consequence, small errors in its determination result in large uncertainty of the calculated spring constant value.

In case of triangular cantilevers, calculations are far more complex. One can estimate the spring constant of these cantilevers approximating the shape as a doubled rectangular cantilever. However, such approach does not precisely estimate the elastic properties of the cantilever. The most complete analytical description of the cantilever bending was proposed by Neumeister et al. [16]. According to the presented analytical calculations, the shape of the cantilever can be approximated as a triangular plate and two identical prismatic beams (Fig. 3.13b). In this manner, the bending of the whole cantilever upon the normal F_N can be split into two movements: bending of a triangular plate and deflection of a single prismatic beam. Thus, the total deflection of the cantilever Δ_N can be written as

$$\Delta_N = \Delta_I + \Delta_{II} + g \cdot \Theta_{II}, \tag{3.13}$$

where Δ_I is the deflection of clamped triangular plate, Δ_{II} is the deflection of two beams and Θ_{II} is their rotation, and the factor g defined as

$$g = \frac{w}{\sin\alpha} - d, \tag{3.14}$$

which depends on the cantilever geometry, i.e., the width of the beam and the opening angle (d is the position of the tip center). The corresponding equations of the deflections and rotation are as follows:

$$\Delta_I = \frac{3 \cdot F}{E \cdot t^3 \cdot \tan a} \cdot \left\{ \left(\frac{w}{\sin\alpha} - 2 \cdot d \right)^2 - d^2 \cdot \left(2 \cdot \log\left[\frac{w}{d \cdot \sin\alpha} \right] + 1 \right) \right\} \tag{3.15}$$

$$\Delta_{II} = \frac{F \cdot L^2}{E \cdot w \cdot t^3 \cdot \cos^2\alpha} \cdot \left\{ \left(\frac{2 \cdot L}{\cos\alpha} + 3 \cdot (w \cdot \cot\alpha - d \cdot \cos\alpha - r \cdot \sin\alpha) \right) \right\} \tag{3.16}$$

$$\Theta_{II} = \frac{2 \cdot F \cdot L \cdot (1 + \nu)}{E \cdot w \cdot t^3 \cdot \cos\alpha} \cdot \left(\frac{w}{\sin\alpha} - d + r \cdot \cot\alpha \right), \tag{3.17}$$

where ν is the Poisson's coefficient (describing the compressibility of the material) and r is the length given by

$$r = \frac{L \cdot \tan\alpha + (w - d \cdot \sin\alpha) \cdot (1 - \nu) \cdot \cos\alpha}{2 - (1 - \mathrm{n}) \cdot \cos^2 \alpha} \tag{3.18}$$

On the other hand, the Eq. (3.13) can be rewritten as

$$\Delta_{\mathrm{N}} = \frac{F_{\mathrm{N}}}{k_{\mathrm{cant}}}, \tag{3.19}$$

where k_{cant} is the normal stiffness of the cantilever, and it can be calculated as follows:

$$k_{\mathrm{cant}} = \frac{F_{\mathrm{N}}}{\Delta_{\mathrm{I}} + \Delta_{\mathrm{II}} + \left(\frac{w}{\sin\alpha} - d \right) \cdot \Theta_{\mathrm{II}}} \tag{3.20}$$

assuming that the analytical forms of Δ_{I}, Δ_{II}, and Θ_{II} are known.

Thermal excitations of the cantilever

The main problem of the abovementioned calculations is a lack of the exact knowledge of the cantilever thickness and Young's modulus value of the material. Therefore, in order to avoid these calculations, several methods for the spring constant calibration have been elaborated. Among them, the most popular one uses the measurement of the resonant frequency (ω_{meas}) of thermally excited cantilever [17, 18]. In the most convenient way, one assumes that the actual mass of the cantilever is equal to the nominal one. Then, the following equation can be applied:

$$k_{\mathrm{cant}} = \frac{k_{\mathrm{N}} \cdot \omega_{\mathrm{meas}}^2}{\omega_{\mathrm{N}}^2}, \tag{3.21}$$

where ω_{N} and k_{N} are the nominal values of the thermally excited resonant frequency and the spring constant of the cantilever, respectively. The example of the resonant frequency spectrum obtained on the basis of the noise measurement of the thermally excited cantilever and the applied fast Fourier transform (FFT) is presented in Fig. 3.14.

The measurements of resonance frequency result in relatively narrow frequency distributions as presented in Fig. 3.15. The histograms were obtained for two types of cantilevers (MLCT type C and D) coming from the same silicon wafer (i.e., undergoing the same production conditions).

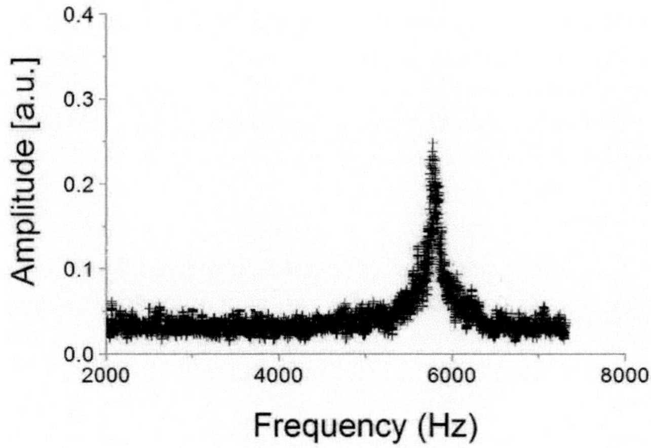

Figure 3.14 Resonant frequency measured for the thermally excited cantilever. The obtained resonant frequency is 5.8 kHz, which corresponds to the calculated cantilever spring constant of 0.007 N/m. Reprinted with permission from [9].

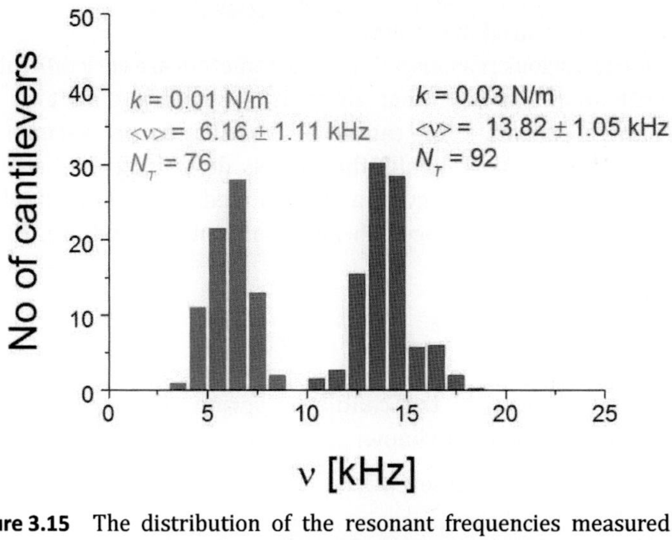

Figure 3.15 The distribution of the resonant frequencies measured for two types of cantilevers with nominal spring constants of 0.01 N/m and 0.03 N/m originating from the same wafer and used later on for the AFM measurements. NT is the number of measured cantilevers. Reprinted with permission from [9].

The maxima, corresponding to cantilevers with the nominal spring constants k_N of 0.01 N/m and 0.03 N/m, were centered at 6.16 kHz and 13.82 kHz, while the nominal resonant frequencies ω_N given by the manufacturer were 7 kHz and 15 kHz. The obtained frequency values were smaller than the nominal ones by about 10% for both cantilevers. In consequence, cantilever spring constants determined from the Eq. (3.21) were of 0.008 N/m and 0.026 N/m, as compared respectively to 0.01 N/m and 0.03 N/m given by the manufacturer.

Sader method

The method for cantilever calibration proposed by Sader et al. relies on the measurements of a resonant frequency of unloaded cantilever, its dimensions, knowledge of its density and mass of the surrounding medium [17]. The basic equation for the spring constant *k* is

$$k_{\text{cant}} = M_{\text{e}} \cdot \rho_{\text{cant}} \cdot w \cdot t \cdot L \cdot \omega_{\text{vac}}^2, \tag{3.22}$$

where M_e is the normalized effective mass of the cantilever, ρ_{cant} is the density of the cantilever, w, t, L are cantilever's width, thickness, and length, respectively, and ω_{vac} is the resonant frequency measured in vacuum.

Due to various reasons, all these parameters are very difficult to be estimated. Knowing that there is a significant shift in the resonant frequencies of cantilevers oscillating in vacuum as compared to air or to liquid, the models describing the behavior of the cantilever frequency in a viscous medium must be applied. When a cantilever is placed in air or any other fluid, Eq. (3.22) takes the following form:

$$k_{\text{cant}} = 0.1906 \cdot \rho_{\text{f}} \cdot w^2 \cdot L \cdot Q_{\text{f}} \cdot \Gamma_{\text{i}}(\omega_{\text{f}}) \cdot \omega_{\text{f}}^2 \tag{3.23}$$

Thus, to calculate the cantilever spring constant using the Sader method, one has to know:

(1) length of the cantilever, L;
(2) width of the cantilever, w;
(3) resonant frequency of the cantilever, ω_f;
(4) quality factor of the resonant curve, Q_f;

(5) density of the fluid surrounding the cantilever, ρ_f;

(6) viscosity of the fluid surrounding the cantilever, that is described by the imaginary part of the (dimensionless) hydrodynamic function evaluated at the resonant frequency, $\Gamma_i(\omega_f)$.

Thermal tune method

The basis for the thermal tune approach [19] lies in the equipartition theorem, which states that for every degree of freedom of any physical system the energy is equal to the energy stored in the corresponding spring:

$$\frac{1}{2}\cdot k_{\text{meas}}\cdot\left\langle z^2\right\rangle=\frac{1}{2}\cdot k_{\text{B}}\cdot T, \tag{3.24}$$

where k_B is the Boltzmann's constant, T is the temperature, $\langle z^2\rangle$ is the mean square displacement denoting deflections of the cantilever characterized by a spring constant k_{meas}. Assuming that the cantilever is an ideal harmonic oscillator, the spring constant can be determined from the time-dependent measurements of the thermal noise. Values of the force constant of the cantilever are calculated using the mean square displacement that can be determined either by averaging over the deflection signal measured as a function of time or by integrating over $\langle z^2\rangle$ in the frequency domain (in the other words, by integrating the power spectral density):

$$k_{\text{meas}}=\frac{k_{\text{B}}\cdot T}{P}, \tag{3.25}$$

where P is the area of the power spectral density function of the thermal fluctuations.

The thermal tune method accounts for the cantilever stiffness in the direction of piezoscanner movement that should be perpendicular to the oscillating cantilever. Therefore, several correction factors should be considered such as tilt angle, cantilever tip position and length.

The tilt of the cantilever (C_{tilt}) can be corrected using the following formula:

$$C_{\text{tilt}}=\frac{1}{\cos^2\theta}, \tag{3.26}$$

where θ is the angle of the cantilever tilt that is usually 10–12°.

The cantilever stiffness depends on the position of the tip, which is rarely positioned exactly at the end of the cantilever. For a rectangular cantilever, stiffness is a function of the length cubed, therefore, the position of the tip follows the relation:

$$C_{\mathrm{pos}} = \frac{L^3}{L_{\mathrm{tip}}^3}, \tag{3.27}$$

where C_{pos} is the correction factor for the cantilever position, L is the length of the cantilever and L_{tip} is the distance from the base of the cantilever to the tip location.

The length of the cantilever probe tip (C_L) also impacts the loading of a cantilever as described by Hutter [20]. The correction factor to account for the probe tip height, h, is

$$C_L = \left(\frac{1 - \frac{2 \cdot h}{L} \cdot \tan\theta}{1 - \frac{3 \cdot h}{L} \cdot \tan\theta} \right)^2, \tag{3.28}$$

Thus, the final calibration formula is:

$$k_{\mathrm{cant}} = \frac{k_{\mathrm{B}} \cdot T}{\mathrm{invPSD}^2} \cdot \frac{P}{\cos^2\theta} \cdot \frac{L^3}{L_{\mathrm{tip}}^3} \cdot \left(\frac{1 - \frac{2 \cdot h}{L} \cdot \tan\theta}{1 - \frac{3 \cdot h}{L} \cdot \tan\theta} \right)^2, \tag{3.29}$$

where invPSD is the inverse of the photodetector sensitivity (expressed in m/V), P is the positional noise power (in V^2) isolated in the fundamental resonant mode only.

Added mass method

A detailed, more precise procedure of the spring constant determination has been proposed by Cleveland et al. [21]. In this method, the cantilever spring constant is determined from the shift in resonance frequency resulting from the presence of known mass placed at the free end of the cantilever (Fig. 3.16):

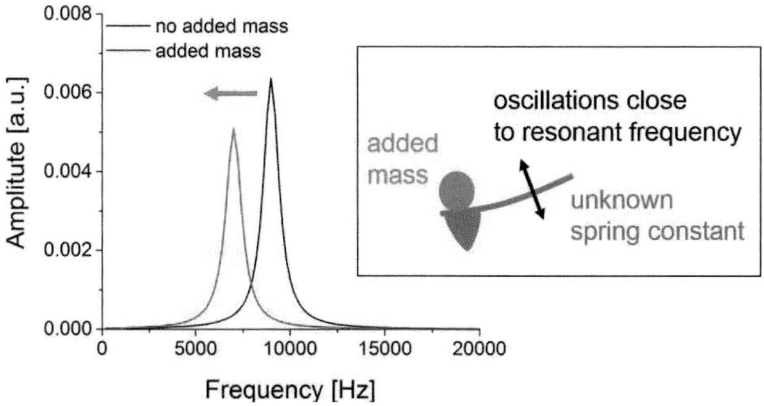

Figure 3.16 The idea of the added mass methods. A particle with a known mass is placed at the free end of the cantilever that oscillates at its resonance frequency. The presence of mass leads to frequency shift.

The presence of added mass shifts the position of the resonance peak to lower values. The resonant frequency depends on mass in a following way:

$$\omega_{\text{res}} \approx \frac{k_{\text{meas}}}{m_{\text{eff}}}, \tag{3.30}$$

where ω_{res} is the angular resonant frequency, k_{meas} is the cantilever spring constant, and m_{eff} is the effective mass of the cantilever. With the added mass m_{a} attached, the resonant frequency shift is described by the following equation:

$$\omega_{\text{res}}^2 \approx \frac{k_{\text{meas}}}{m_{\text{eff}} + m_{\text{a}}} \tag{3.31}$$

Then

$$m_{\text{a}} = \frac{k_{\text{meas}}}{\omega_{\text{meas}}^2} - m_{\text{eff}}. \tag{3.32}$$

By measuring resonance frequency as a function of mass added to the cantilever, a plot of m_{a} versus $1/\omega_{\text{meas}}^2$ is a straight line with a slope directly delivering cantilever spring constant and with an intercept representing the effective mass of cantilever m_{eff}. The mass of the cantilever can be calculated by multiplying the density of the silicon nitride (2.8 g/cm^2) by the volume of the cantilever (determine from its geometrical shape).

This method is dependent on the position where added mass is placed and thus needs the correction factor. When the mass is placed at a distance ΔL away from the tip along the length L of the cantilever, the calculated spring constant value k_m is related to the spring constant k_{meas} determined for the correct position by

$$k_{meas} = k_m \left(\frac{L - \Delta L}{L} \right)^3. \tag{3.33}$$

The mass correction is

$$m_a = m_{eff} \left(\frac{L - \Delta L}{L} \right)^3. \tag{3.34}$$

Both relations (3.33) and (3.34) show that a small discrepancy in the position where mass is added introduces a cubic error.

Method of comparing with the known spring constant

Analogously to thermal excitation methods, cantilever spring constant can be determined by comparison with the cantilever with a known spring constant [22], attached to a holder (Fig. 3.17).

The reference cantilever, with known spring constant k_{ref}, is mounted directly on a piezoelectric scanner while the unknown one is mounted at the fixed position (cf. Fig. 3.17). Laser beam is focused on the fixed cantilever. Then, in normal scanner approach–retract operation, a force curve is recorded. From these data, deflection δ_{meas} of the unknown cantilever may be determined, following the position of the piezoelectric scanner, δ_{ref}. The latter value corresponds to the deflection of the reference cantilever. Knowing the deflection of the unknown cantilever δ_{meas} and the height of the piezoelectric scanner δ_{ref}, the spring constant of unknown cantilever k_{meas} is given by

$$k_{meas} = k_{ref} \cdot \frac{\delta_{ref} - \delta_{meas}}{\delta_{meas}}. \tag{3.35}$$

The zero deflection is assumed for the moment when the tip just touches the reference cantilever. In fact, the spring constant k_{meas}

can be measured from the slope of the force curve ($\delta_{\text{meas}}/\delta_{\text{ref}}$) obtained on the reference cantilever recorded in contact mode, according to

$$k_{\text{meas}} = k_{\text{ref}} \cdot \frac{1 - \frac{\delta_{\text{meas}}}{\delta_{\text{ref}}}}{\frac{\delta_{\text{meas}}}{\delta_{\text{ref}}}}. \tag{3.36}$$

A reference cantilever should have a spring constant close to the one that will be calibrated. In this method, the reference cantilever should be characterized by a well-defined spring constant value, what is not always easy to assure. Additionally, the surface contamination and the place of contact between two cantilevers (the position whether the tip of unknown cantilever touches the surface of the reference one) influences the determination of the inverse of the photodetector sensitivity value.

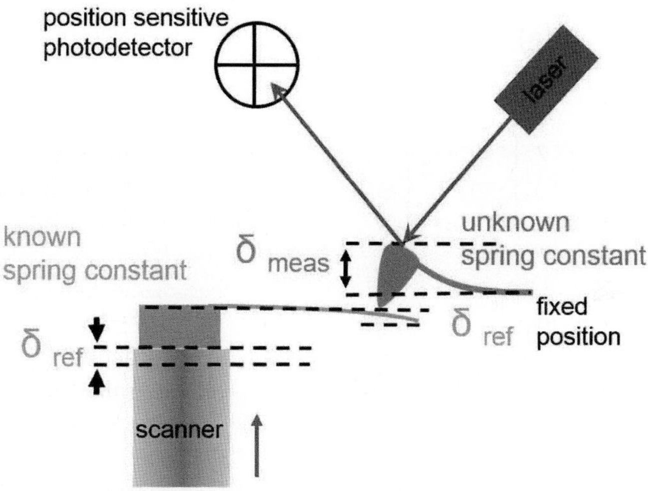

Figure 3.17 The idea of cantilever spring calibration based on deflection measurements of a cantilever with known spring constant (e.g., standard, tipples one).

3.2.1.4 Force versus sample-distance conversion

In the AFM, usually, a sample (or an AFM tip) is moved along *Z*-axis. Thus, the relative scanner (or sample) position recorded during the AFM measurement is not a distance between a tip and a sample [9]. The distance between the sample surface and the probing cantilever, referred here as a displacement D, can be calculated by subtracting the cantilever deflection at the "rest" position (d_C) from the relative position of the piezoelectric scanner (Z) as it is schematically shown in Fig. 3.18a for the case of hard, non-deformable surface. If the sample is non-deformable, the tip–sample distance, D, is the difference between Z scanner position and cantilever deflection:

$$D = Z - d_c \tag{3.37}$$

If the sample can be deformed, the resulting cantilever deflection will be smaller as compared to a hard sample. The tip-sample distance considers the deformation depth (Fig. 3.18b):

$$D = Z - (d_c + d_s), \tag{3.38}$$

where d_S is the depth of the sample deformation.

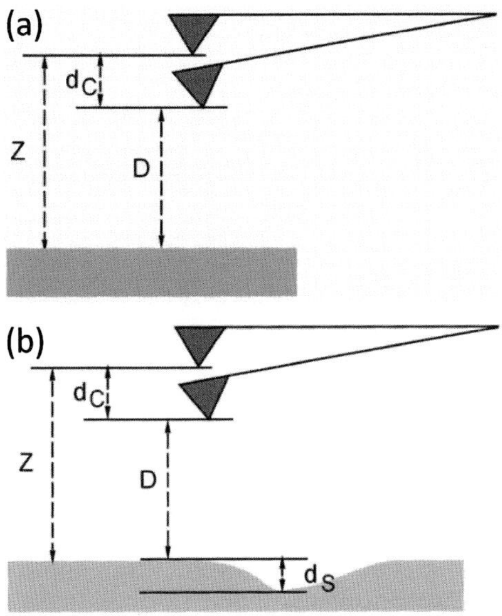

Figure 3.18 The tip-sample distance D determination on (a) a stiff, non-deformable surface and (b) in case of soft sample where the depth of the sample deformation d_S should be taken into account. (Z is the relative piezoelectric scanner position), d_C is the "zero" position of the cantilever. Reprinted with permission from [9].

3.2.1.5 Hydrodynamic drag

Like any object moving in a liquid, the cantilever experiences a force arising due to viscous friction with the surrounding medium [23–25], i.e., so-called hydrodynamic drag force, F_{drag}, that acts always in the opposite direction to the cantilever movement (Fig. 3.19).

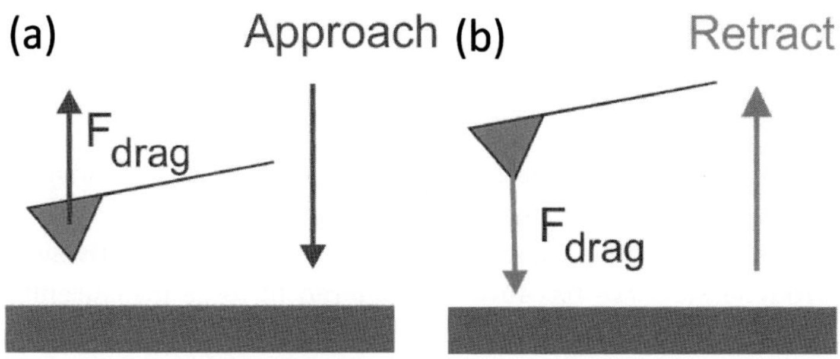

Figure 3.19 The hydrodynamic drag forces acting on AFM cantilever during (a) approach and (b) retract. Reprinted with permission from [9].

Thus, the force measured by AFM ($F = k_{cant} \cdot z$) is a sum of the true acting force (F_{true}) and the hydrodynamic force (F_{drag}). Its value is smaller when cantilever approaches the surface:

$$k_{cant} \cdot z = F_{true} - F_{drag} \tag{3.39}$$

and it is larger when the cantilever is withdrawn from the surface:

$$k_{cant} \cdot z = F_{true} + F_{drag} \tag{3.40}$$

As it has been shown by Alcaraz et al. [24], the drag force acting on the cantilever can be described by the following equation, valid for low Reynolds number $Re < 1$:

$$F_{drag} = \frac{6 \cdot \pi \cdot \eta \cdot a_{eff}^2}{h + h_{eff}} \cdot v_{eff}, \tag{3.41}$$

where η is the viscosity of the liquid, h is the tip-surface separation, a_{eff} is the effective tip radius, h_{eff} is the effective tip height, and v_{cant} tip denotes the velocity of the AFM cantilever. Currently, there is no satisfactory theoretical model describing the dependence of the drag force on the cantilever geometry. Despite that, it has been shown that the h_{eff} value is close to the nominal value of the tip height (~3 μm), and the effective tip radius a_{eff} is of about 22 μm.

The hydrodynamic force increases with the velocity of the approaching and retracting tip. In almost all recent AFM measurements, the tip velocity varied from few nm/s to about 10 μm/s [25]. At first approach, all biological buffers can be treated as water since usually they are aqueous solutions. Thus, assuming the viscosity of surrounding liquid to be close to the water viscosity, the estimated maximum value of the hydrodynamic force contributing to the total acting force is 0.06 pN and 6.28 pN for the tip velocity of 1 μm/s and 10 μm/s, respectively. The hydrodynamic force has to be considered in force measurements at large velocities when it approaches the magnitude of the measured interaction force. Additionally, the hydrodynamic force is not constant over the tip-sample distance, leading to the enhancement of the effect at small tip-surface separations.

Experimentally, F_{drag} can be estimated from the separation of the approach and retract curves in their non-contact regions (i.e., base line, Fig. 3.20). Such estimation is similar to the procedure of AFM determination of the friction force [24].

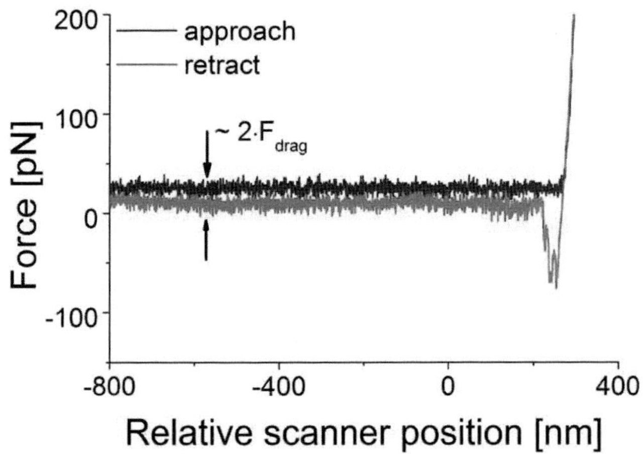

Figure 3.20 The idea of the measurement of hydrodynamic force, F_{drag}, based on a force curve, recorded during approach and retract. Reprinted with permission from [9].

Two exemplary distributions of hydrodynamic forces recorded during moving the cantilever in PBS buffer at two velocities: 2.4 μm/s (a) and at 7.7 μm/s are presented in Fig. 3.21.

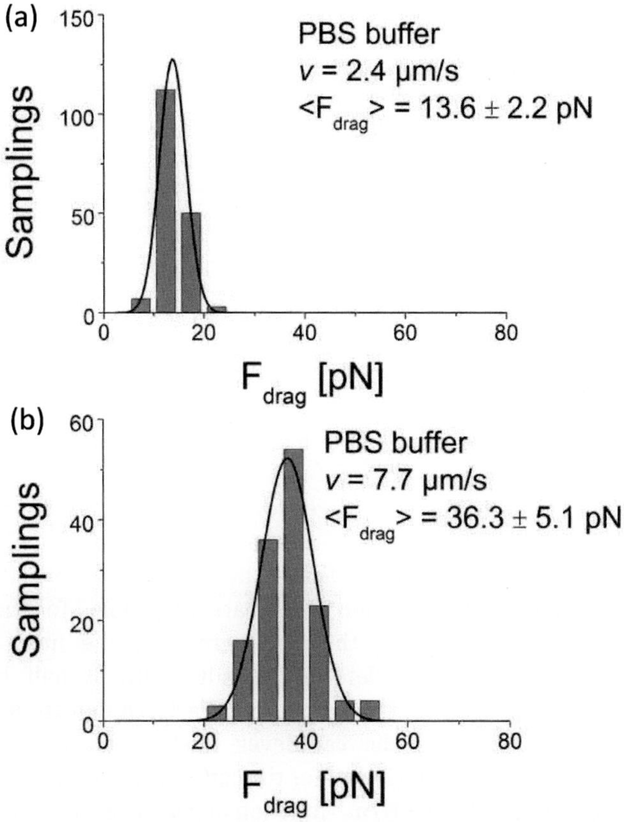

Figure 3.21 Distributions of the hydrodynamic force measured in PBS buffer, at the cantilever velocity of (a) 2.4 μm/s and (b) 7.7 μm/s. Reprinted with permission from [9].

The corresponding hydrodynamic drag forces are 13.6 pN to 36.3 pN. Thus, assuming that the force measured by the AFM is 100 pN, the true acting force is 81.4 pN and 63.7 pN, respectively.

3.2.1.6 Force detection limit

The detection limit value of the force is linked with the cantilever spring constant, k_{cant}, by the following formula [26]:

$$F_{\text{Det}} = \sqrt{k_{\text{cant}} \cdot k_{\text{B}} \cdot T} \tag{3.42}$$

where k_B is the Boltzmann constant, and T is the temperature. Table 3.3 presents the theoretical force detection limit calculated for typical cantilevers.

Table 3.3 Force detection limit of cantilevers typically used for various biological applications, estimated for the room temperature of 23°C

$k = 0.01$ N/m	$F_{\text{Det}} = 6.6$ pN
$k = 0.03$ N/m	$F_{\text{Det}} = 11.4$ pN
$k = 0.1$ N/m	$F_{\text{Det}} = 20.9$ pN
$k = 0.5$ N/m	$F_{\text{Det}} = 46.7$ pN

Depending on the cantilever spring constant, the detection limit changes significantly and inappropriate choice of the cantilever may make impossible measurements of the interaction force between single molecules.

Experimentally, the force detection limit is usually determined from the noise fluctuations of the base line, i.e., the horizontal parts of the force curve recorded far away from the investigated surface, where the interaction forces are negligible. To quantify the cantilever fluctuations, their histogram can be fitted with the Gauss function. The determined full width at half height (FWHH) can be used as an estimate of the force detection limit. Figure 3.22b,d presents the cantilever fluctuations around the base line measured in PBS buffer for the cantilever with the nominal spring constant of 0.01 N/m, functionalized with the antibody against prostate specific membrane antigen. The FWHM value in this particular experiment was ~10 pN. However, in similar experiment (Fig. 3.22c,d) even though the same cantilever type was used, the FWHM value was much larger (41.9 pN), indicating much higher noise level.

Apart from thermal excitations, there are many possible sources of the cantilever fluctuations. They may be due to (i) the quality of the surface of the liquid-cell setup, (ii) transparency of the buffer used in experiment, (iii) the fact that the sample or its fragments may detach from the surface and also, (iv) the quality of the procedure of cantilever functionalization.

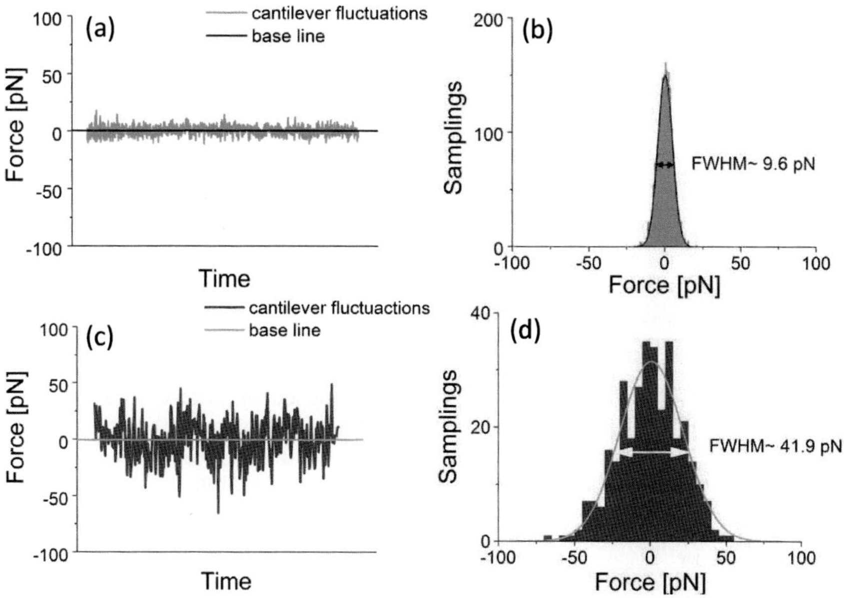

Figure 3.22 (a) and (c) Cantilever fluctuations around the base line measured in PBS buffer for two randomly chosen cantilevers, both with the nominal spring constant value of 0.01 N/m, functionalized with the antibody against prostate specific membrane antigen. (b) and (d) The corresponding histograms of the cantilever fluctuations. FWHM defines the range of fluctuations. Reprinted with permission from [9].

3.2.1.7 Scanner linearization

Since the knowledge of the true scanner displacement is crucial for all calculations, the piezoelectric scanner nonlinearity and hysteresis should be carefully examined. In commercially available AFMs, scanner linearization is frequently an integral part of the hardware. A simple way to verify the precision of the scanner performance is to measure a force versus relative sample position curve on a stiff, non-deformable surface and to verify whether the approach curve is identical (overlaps) to the retraction one and if it is represented by a straight line. If it is true, the piezoelectric scanner is hardware-linearized. Otherwise, the piezoelectric scanner should be linearized off-line. Among several methods of scanner linearization, the method of z-calibration proposed by Jaschke and Butt [27] is straightforward and does not require any additional equipment (Fig. 3.23a).

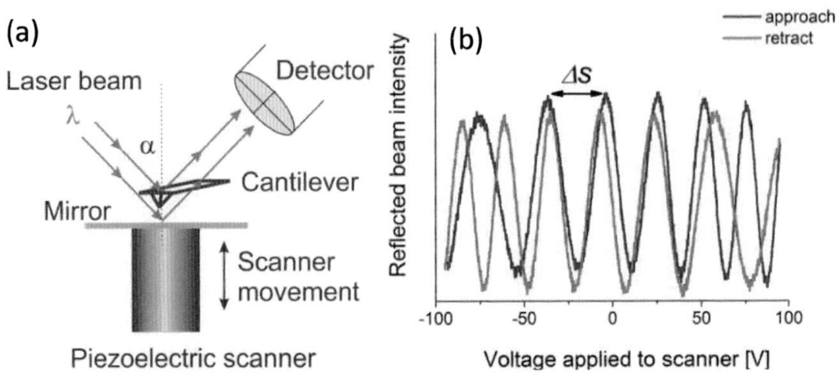

Figure 3.23 (a) The idea of the method proposed by Jaschke and Butt [23] for the piezoelectric scanner calibration. (b) Reflected beam intensity as a function of the voltage applied to the scanner recorded in both direction of scanner movement: to and from the mirror. The distance (Δs) is between two subsequent interference maxima. Reprinted with permission from [9].

It can be applied only to AFMs equipped in an optical detection system. In this method, a mirror is placed in the sample position in such a way that the inclination of a mirror is the same as the one of a cantilever. Slightly unfocused laser beam produces two reflected beams: from the back of cantilever and from the mirror surface. When the mirror moves up and down, the optical path lengths of two beam fractions change and the resulting interference signal varies (Fig. 3.23b). The measured distances between subsequent maxima reflect the scanner nonlinearity.

Using the Bragg law, it is possible to calculate the true scanner extension:

$$\Delta s = \frac{n \cdot \lambda}{2 \cdot \sin \alpha}, \tag{3.43}$$

where λ is the wavelength of the laser beam, α is the angle of the incident beam and n is the number of interference maxima. In case of the linearized scanner, the two traces of the reflected intensity, recorded during the up and down movement of the scanner, should overlap. If not, the relation between the determined positions of the maxima and the applied voltage should be quantified. The scanner nonlinearity and hysteresis can be described by a polynomial function—usually a quadratic approximation is sufficient:

$$\text{True extension} = H_{\text{Q}} \cdot (\text{voltage})^2 + H_{\text{L}} \cdot (\text{voltage}), \tag{3.44}$$

where H_{Q} and H_{L} are quadratic and linear coefficients.

In commercial devices, the scanner linearization is usually performed automatically without any action of the user. However, taking into account the ageing of piezoelectric materials, the linearity of scanner should be measured from time to time since the incorrect parameters influence not only the true scanner displacement but also the photodetector sensitivity used for the force determination.

3.2.1.8 Scanner velocity determination

The retraction velocity (describing how fast two molecules are separated) is essential for the determination of the loading rate value, since the measured unbinding force depends on how fast the applied rupture force changes in time [28]. Since, in AFMs, the separation is realized by the movement of the piezoelectric scanner, its nonlinearity influences strongly the values of the retraction velocity (Fig. 3.24).

Only if a hardware linearization is provided, the force change in time can be assumed linear. How strongly the scanner non-linearity influences the determination of the retraction velocity can be demonstrated for the apparatus working at the IFJ PAN. It operates using a non-linearized piezoelectric scanner.

During retraction from point A to B, the temporary speed calculated from the tangential significantly decreases with time. Therefore, simple approximation of the scanner retraction velocity introduces inaccuracy in the loading rate values [28]. The use of exact, variable value of retraction velocity in calculations can vastly improve the quality of the unbinding force histograms as observed, for example, for the unbinding force between prostate specific membrane antigen (PSMA) and monoclonal antibody against PSMA (Fig. 3.25a,b).

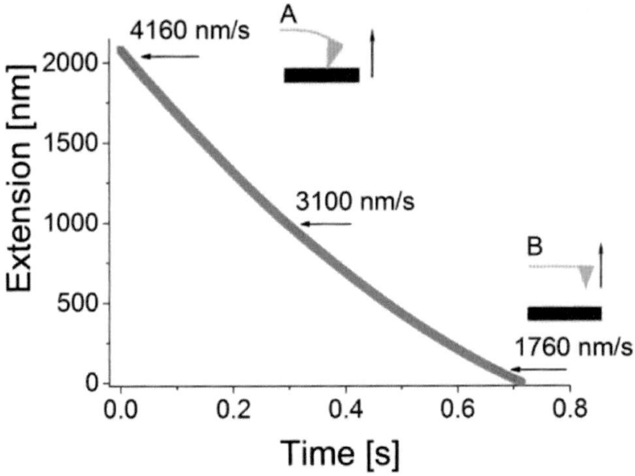

Figure 3.24 Relation between the scanner elongation and time determined for a scanner type EBL #2 (from Staveley Sensors Inc.). Arrows indicate the temporary speed. Reprinted with permission from [9].

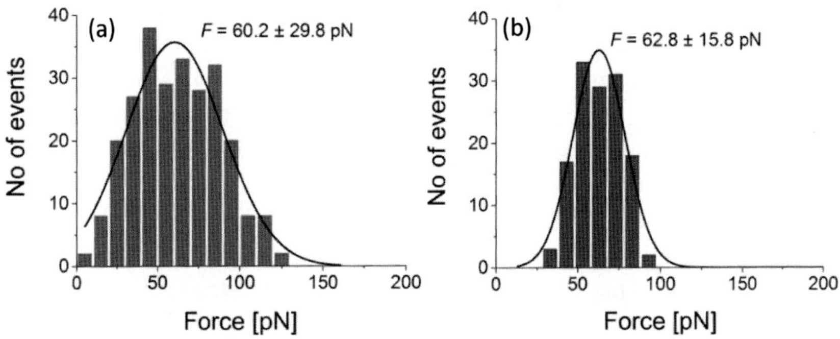

Figure 3.25 Force histograms determined for the unbinding of prostate specific membrane antigen (PMSA) and its monoclonal antibody, before (a) and after (b) exact determination of the retraction velocity. The same set of data is shown on both histograms. Unpublished data of the author.

The effect of retraction velocity is twofold. Firstly, it manifests in a slight shift of the force distribution center (the relative change of the unbinding force was of about 4%). Such error lies within the experimental uncertainties originating from the noise level observed in the AFM experiments. The larger effect is visible in the distribution width. After exact determination of the retraction velocity, the width of the force histogram decreased almost two times.

The retraction velocity is not the only factor influencing the width of the force histogram. The other factors are, for example, the noise level (and thereby the force detection limit) and the physics of the unbinding process (i.e., the dependence on the loading rates).

References

1. Binnig, G., Quate, C. F., and Gerber, C. (1986). Atomic force microscopy. *Phys. Rev. Lett.*, **56**, pp. 930–933.
2. Bouckaert, J., Poortmans, F., Wyns, L., and Loris, R. (1996). Sequential structural changes upon zinc and calcium binding to metal-free concanavalin A. *J. Biol. Chem.*, **271**, pp. 16144–16150.
3. Engel, A., Schoenenberger, C.-A., and Müller, D. J. (1997). High resolution imaging of native biological samples using scanning probe microscopy. *Curr. Opin. Struct. Biol.*, **7**, pp. 279–284.
4. Chen, C., and Hamers, R. (1991). Role of atomic force in tunneling–barrier measurements. *J. Vac. Sci. Technol. B*, **9**, pp. 503–505.
5. Hipp, M., Bielefelt, H., Colchero, J., Marti, O., and Mlynek, J. (1992). A stand-alone scanning force microscope. *Ultramicroscopy*, **42–44**, pp. 1498–1503.
6. Hues, S. M., Drapper, C. F., Lee, K. P., and Colton, R. J. (1994). Effect of PTZ and PMN actuator hysteresis and creep on nanoindentation measurements using force microscopic tips. *Rev. Sci. Instrum.*, **64**, pp. 1561–1565.
7. Butt, H. J., Capella, B., and Kappl, M. (2005). Force measurements with the atomic force microscope: Technique, interpretation and applications. *Surf. Sci. Rep.*, **59**, pp. 1–152.
8. Cappella, G., and Dietler, G. (1999). Force-distance curves by atomic force microscopy. *Surf. Sci. Rep.*, **34**, pp. 1–104.
9. Lekka, M. (2007). The use of atomic force microscopy as a technique for the identification of cancerous cells. *Report IFJ PAN*, 2001/AP, pp. 1–147.
10. Lekka, M., Laidler, P., Gil, D., Lekki, J., Stachura, Z., and Hrynkiewicz, A. Z. (1999). Elasticity of normal and cancerous human bladder cells studied by scanning force microscopy. *Eur. Biophys. J.*, **28**, pp. 312–316.

11. Weisenhorn, A., Khorsandi, M., Kasas, S., Gotzos, V., and Butt, H. J. (1993). Deformation and height anomaly of soft surfaces studied with an AFM. *Nanotechnology*, **4**, pp. 106–113.
12. Sneddon, I. N. (1965). The relation between load and penetration in the axisymmetric Boussinesq problem for a punch of arbitrary profile. *Int. J. Eng. Sci.*, **3** pp. 47–57.
13. Butt, H.-J., and Jaschke, M. (1995). Calculation of thermal noise in atomic force microscopy. *Nanotechnology*, **6**, pp. 1–7.
14. Proksch, R., Schäffer, T. E., Cleveland, J. P., Callahan, R. C., and Viani, M. B. (2004). Finite optical spot size and position corrections in the spring constant calibration. *Nanotechnology*, **4**, pp. 1344–1350.
15. Ohler, B. (2007). Cantilever spring constant calibration using laser Doppler vibrometry. *Rev Sci. Instrum.*, **78**, p. 063701.
16. Neumeister, J. M., and Ducker, W. A. (1994). Lateral, normal, and longitudinal spring constants of atomic force microscopy cantilevers. *Rev. Sci. Instrum.*, **65**, pp. 2527–2531.
17. Sader, J. E., Larson, I., Mulvaney, P., and White, L. R. (1995). Method for the calibration of atomic force microscope cantilevers. *Rev. Sci. Instrum.*, **66**, pp. 3789–3798.
18. Green, C. P., Lioe, H., Cleveland, J. P., Proksch, R., Mulvaney, P., and Sader, J. E. (2004). Normal and torsional spring constants of atomic force microscope cantilevers. *Rev. Sci. Instrum.*, **75**, pp. 1988–1996.
19. Hutter, J. L., and Bechhoefer, J. (1993). Calibration of atomic-force microscope tips. *Rev. Sci. Instrum.*, **64**, pp. 1868–1873.
20. Hutter, J. L. (2005). Comment on tilt of atomic force microscope cantilevers: Effect on spring constant and adhesion measurements. *Langmuir*, **21**, pp. 2630–2632.
21. Cleveland, J. P., Manne, S., Bocek, D., and Hansma, P. K. (1993). A nondestructive method for determining the spring constant of cantilevers for scanning force microscopy. *Rev. Sci. Instrum.*, **64**, pp. 403–405.
22. Kim, M. S., Choi, J. H., Kim, J. H., and Park, Y. K. (2010). Accurate determination of spring constant of atomic force microscope cantilevers and comparison with other methods. *Measurement*, **43**, pp. 520–526.
23. Janoviak, H., Struckmeier, J., and Muller, D. J. (2005). Hydrodynamic effects in fats AFM single-molecule force measurements. *Eur. Biophys. J.*, **34**, pp. 91–96.

24. Koleske, D. D. Lee, G. U., Gans, B. I., Lee, K. P., DiLella, D. P., Wahl, K. J., Barger, W. R., Whitman, L. J., and Colton, R. J. (1995). Design and calibration of a scanning force microscope for friction, adhesion, and contact potential studies. *Rev. Sci. Instrum.*, **66**, pp. 4566–4574.

25. Alcaraz, J., Buscemi, L., Puig de Morales, M., Colchero, J., Baro, A., and Navajas, D. (2002). Correction of microrheological measurements of soft samples with atomic force microscopy for the hydrodynamic drag on the cantilever. *Langmuir*, **18**, pp. 3578–3588.

26. Gittes, F., and Schmidt, C. (1998). Thermal noise limitations on micromechanical experiments. *Eur. Biophys. J.*, **27**, pp. 75–81.

27. Jaschke, M., and Butt, H. J. (1995). Height calibration of optical lever atomic force microscopes by simple laser interferometry. *Rev. Sci. Instrum.*, **66**, pp. 1258–1259.

28. Evans, E. (2001). Probing the relation between force-lifetime and chemistry in single molecular bonds. *Annu. Rev. Biophys. Biomol. Struct.*, **30**, pp. 105–128.

Chapter 4

Quantification of Cellular Elasticity

During the past decades, a number of studies have proved the link between mechanical properties and cellular functions, emphasizing alterations in cellular mechanical properties correlated with development of various diseases, in particular, in cancer. Recent progress in cancer cell motility and invasion research has resulted in a greater understanding of the role of mechanical properties in malignant transformation. There is evidence that cancer progression is characterized by disruption and/or reorganization of cytoskeleton (i.e., a cellular scaffold [1]). This is accompanied by various molecular alterations influencing the mechanical properties of cells [2].

4.1 Materials Properties and Theoretical Models

4.1.1 Basic Terms Used in Material Mechanics

Quantitative (or even qualitative) characterization of mechanical properties of various materials is important in many fields of our life. The mechanical properties, such as Young's modulus, the index of plasticity, the ultimate strength, the tensile strength, and the elongation, are the examples of parameters describing material mechanical properties. They are usually derived on a

Cellular Analysis by Atomic Force Microscopy
Malgorzata Lekka

ISBN 978-981-4669-67-2 (Hardcover), 978-1-315-36480-3 (eBook)
www.panstanford.com

basis of stress–strain relationship, where stress is the applied load force per unit area while strain is the ratio between deformation and initial length [3–5]. A scheme of stress–strain relation is presented in Fig. 4.1.

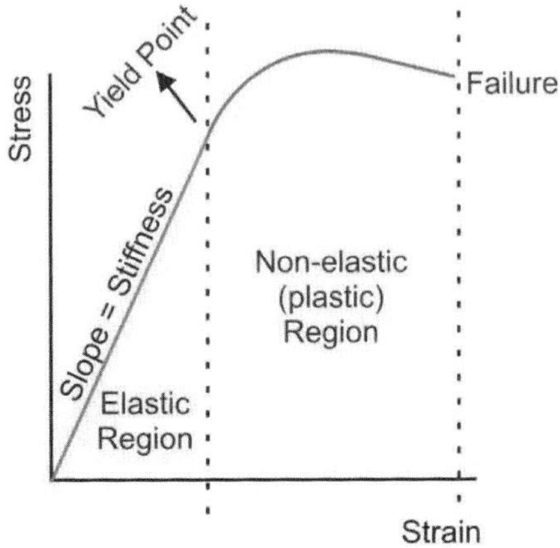

Figure 4.1 Relation between stress and strain for a hypothetical material.

In the low-strain region, many materials obey the Hooke's law saying that a stress σ is proportional to a strain ε ($\sigma = E\cdot \varepsilon$). The proportionality factor is Young's modulus E (frequently called elastic modulus). In this region, materials are ***elastic,*** i.e., if some deformation is produced in a material by applied stress, stress release brings deformation back to zero [4]. A classic example of elastic material is a metal spring. Here, the applied stress changes linearly with a strain following the above-mentioned Hooke's law:

$$F = k \cdot x, \tag{4.1}$$

where k is the elasticity constant of a spring (i.e., spring constant), x is the displacement of the end of the spring from its equilibrium position (i.e., when applied stress equals to zero), and F is the force exerted on the material.

As strain increases, many materials deviate from this linear dependence (material becomes *non-elastic*). The point of transition is termed ***a yield point.*** Once it is reached, the applied stress produces a certain fraction of deformations that is permanent and non-reversible (material becomes ***plastic***). Further strain increase leads to a material *failure.* There are several theories developed to describe plastic material properties (i.e., *material* ***plasticity***) based on three common assumptions:

(a) During plastic deformation, the sample volume does not change.
(b) The directions of principal strains and stresses overlap.
(c) In each point, the maximum shear stress is equal to a specific constant.

The simplest theory describing plasticity considers tensions induced in the material under deformations following the Hooke's law [5]. Lack of proportionality beyond the elastic limit (a yield point) in the stress–strain curve is explained assuming that Young's modulus is not constant. It is expressed as a function of a strain ε in the following manner:

$$E(\varepsilon)=E \quad \text{for} \quad \varepsilon<\varepsilon_l \tag{4.2}$$

$$E(\varepsilon)=E(1-\varphi(\varepsilon)) \quad \text{for} \quad \varepsilon>\varepsilon_l, \tag{4.3}$$

where ε_l is the yield limit, $\varphi = \varphi(\varepsilon)$ describes the plasticity function that is derived from experimental data. This function becomes zero for elastic deformations and increases for the plastic ones. Hence, the common dependence between stress and strain ($\sigma = E\cdot \varepsilon$) can be rewritten into

$$\sigma = E(1-\varphi(\varepsilon)) \tag{4.4}$$

This equation is suitable only for tensions. Any phenomena such as torsion or bending (or both) require more complicated theoretical approaches.

Fluids have characteristic resistance to flow called ***viscosity***, which results from the frictional energy loss [6]. The more viscous fluid induces larger frictional energy losses. Figure 4.2 illustrates schematically the flow of the fluid. The largest flow velocity is on the fluid surface while close to boundaries the velocity drops to zero.

To characterize the material viscoelastic properties, the mechanics of solid bodies (where the Hooke's law is valid and the stress is proportional to the strain) cannot be used. Solid material undergoes only a finite amount of deformations under the applied stress. In fluid mechanics, such approach fails because fluid continuously deforms under an applied shear stress. Thus, here, another law applies, linking shear stress τ with a shear rate, under assumption that velocity profile is linear:

$$\tau = \eta \cdot \frac{dy}{dt}, \tag{4.5}$$

where dy/dt defines the shear rate and the proportionality factor η is the viscosity coefficient. The fluids fulfilling Eq. (4.5) are called *Newtonian liquids.* For *a non-Newtonian liquid,* the shear stress–strain rate relation is not linear and, furthermore, viscosity depends upon the applied shear stress [7].

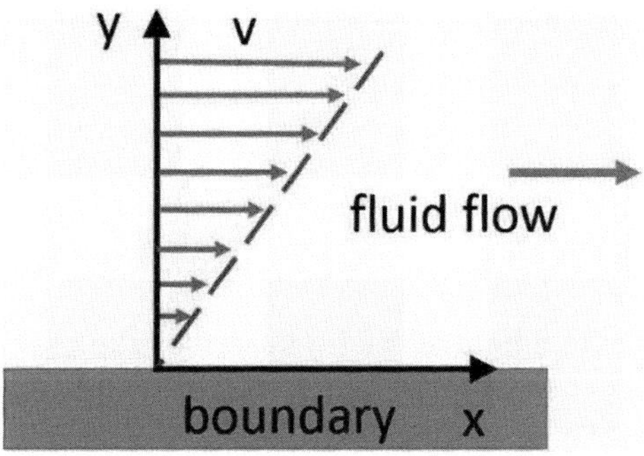

Figure 4.2 Fluid velocity profile showing linear dependence as a function of distance from a boundary (e.g., fixed, bottom plate).

Some materials, such as polymers or cells, behave as both viscous fluids and elastic solids, i.e., they are ***viscoelastic***. Their distinguishable feature is the dependence of mechanical properties on time [8, 9].

4.1.2 Rheological Models

Rheology usually describes the behavior of materials whose mechanical properties are characterized by both elastic ad viscous components, like most of biological objects. However, rheology does not always describe their mechanical properties (e.g., elastic modulus) due to their high structural complexity.

Focusing on mechanical properties of a single cell introduces more complications in the theoretical description of cellular mechanics. Various attempts to model single cell have been reported so far. Some of these models treat a cell as a uniform object, composed of fluid surrounded by elastic membrane, while the others consider also the structural elements present within the cell interior [9–11].

4.1.2.1 Mechanical behavior of soft materials

Mechanical models applied to soft materials (also to a biological cell) must combine both elastic and viscous components [12]. For example, the simplest two-compartment model takes into account an elastic cortical membrane and a viscoelastic cytoplasm [11]. At small deformations, such soft materials behave as the Hook elastic body (*modeled as a spring* for which a dependence between a stress σ and a strain ε is linear) and as the Newtonian fluid (*represented as a dashpot* where stress σ is linearly proportional to time derivative of a strain ε). The equations describing the behavior of such body are as follows [13]:

The Hookean spring: $\sigma = E \cdot \varepsilon$ (4.6)

The Newtonian liquid: $\sigma = \eta \cdot \dfrac{d\varepsilon}{dt}$, (4.7)

where E is the modulus of elasticity, η is the viscosity, and t is the time. The combination of these elements gives a simple description of viscoelasticity.

The simplest rheological models (the Maxwell, Voigt, and Kelvin ones) used to describe viscoelastic properties of living cells are presented in Fig. 4.3.

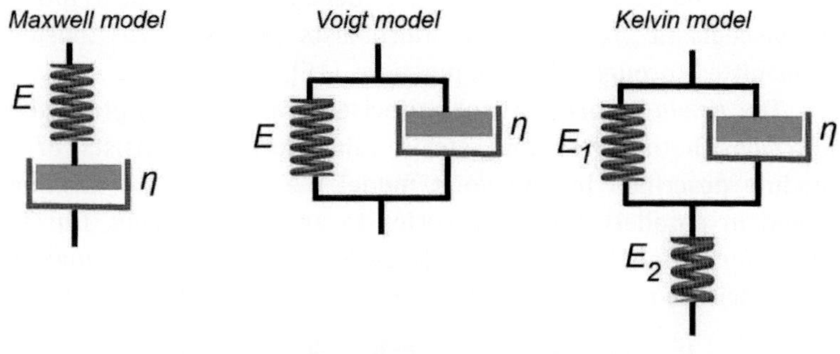

Figure 4.3 Schematic representations of the simplest rheological models, composed of a spring (an elastic component) and a dashpot (a viscous element).

The Maxwell model consists of a viscous element (the Newtonian liquid) linked in series with an elastic element (the Hookean spring). The basic equation for the model is

$$\sigma + \frac{\eta}{E} \cdot \frac{d\sigma}{dt} = \eta \cdot \frac{d\varepsilon}{dt}, \tag{4.8}$$

where E is the modulus of elasticity, η is the viscosity, σ is the stress, ε is the strain, and t is the time.

When such system is quickly loaded, the viscous element is too slow to react and only the spring responses to deformation. On the contrary, if the load is applied slowly, the deformation of the viscous element starts to dominate and interfere with the deformation of a spring. In cellular biology, the Maxwell model is often used to describe the deformation of neutrophils using the micropipette aspiration technique [14].

The Voigt model describes the material consisting of the elastic spring and the viscous dashpot elements connected in parallel (assuming that there is no bending in such arrangement). The strain experienced by the spring is the same as that experienced by the dashpot. In this model, the stress σ depends on a strain ε according to the followingrelation:

$$\sigma = E \cdot \varepsilon + \eta \cdot \frac{d\varepsilon}{dt}, \tag{4.9}$$

where E is the modulus of elasticity, η is the viscosity, and t is the time.

The good example of the Voigt model application is description of viscoelastic response of fibroblasts exposed to tension transmitted through adherent junctions [15].

The Kelvin model (or more precisely *the Kelvin–Voigt model*) describes both stress and strain relaxations. It consists of a module described by the Voigt model (i.e., spring and dashpot linked in parallel), linked in series to an elastic spring. During the deformation, part of the energy is dissipated. The equation that describes mechanical properties of such material is as follows:

$$\sigma + \frac{\eta}{E_1 + E_2} \cdot \frac{d\sigma}{dt} = \frac{E_1 \cdot E_2}{E_1 + E_2} \cdot \varepsilon + \frac{\eta \cdot E_2}{E_1 + E_2} \cdot \frac{d\varepsilon}{dt}, \tag{4.10}$$

where E_2 is Young's modulus of a spring present in the Voigt model, E_1 is Young's modulus of the spring linked in series. The Kelvin–Voigt model has been used, for example, to model mechanical response of collagen molecules delivering information on both Young's modulus and viscosity [16].

More complex models. The above-mentioned models are the simplest ones. They are not always describing mechanical behavior of biological samples in a fully satisfactory way. Thus, more complex models are employed (a so-called *generalized Maxwell* [17] or *generalized Kelvin* models [13, 18]).

The generalized Kelvin model consists of an isolated spring E_0 and N Kelvin units (E_i, η_i) linked in series, described by the following relation:

$$\sigma_i(t) = (E_i + \mathrm{h}_i \cdot \partial/\partial t) \cdot \varepsilon_i(t) \quad (4.11)$$

The stress at each unit is the same external stress $\sigma(t)$ while the total strain $\varepsilon(t)$ is the sum of internal strains in each element. For N elements, the stress–strain relation is following:

$$\sigma(t) = \frac{\varepsilon(t)}{\frac{1}{E_0} + \left(\sum_{i=1}^{N} \frac{1}{E_i + \eta_i \cdot \partial/\partial t}\right)} \quad (4.12)$$

The specific creep function for the generalized Kelvin model is

$$D(t-\tau) = \frac{1}{E_0} + \sum_{i=1}^{N} \frac{1}{E_i} \cdot \left(1 - e^{-\frac{t-t}{\theta_i}}\right), \quad (4.13)$$

where $\theta_i = \eta_i/E_i$ and τ is the characteristic creep time of a single Kelvin unit.

The generalized Maxwell model consists of N different Maxwell units (E_i, η_i) linked in parallel (it is important to note that the absence of the isolated spring ensures fluid-type behavior).

$$\frac{\partial}{\partial t}\varepsilon(t) = \left(\frac{1}{E_i}\frac{\partial}{\partial t} + \frac{1}{\eta_i}\right) \cdot \sigma_i \quad (4.14)$$

For the generalized Maxwell model the strain is the same for all constituting elements and the total stress is given by the following equation:

$$\sigma(t) = \left(E_\infty + \sum_{i=1}^{N} \frac{\partial/\partial t}{\frac{\partial/\partial t}{E_i} + \frac{1}{\eta_i}} \right) \cdot \varepsilon(t) \tag{4.15}$$

The relaxation function is

$$E(t-\tau) = E_\infty + \sum_{i=1}^{N} E_i \cdot e^{\frac{t-\tau}{T_i}}, \tag{4.16}$$

where $T_i = \eta_i/E_i$ and τ is the characteristic relaxation time for single Maxwell unit.

The generalized models introduce more elements to be considered to fully describe the response of real materials, and, simultaneously, they result in larger number of material parameters to be evaluated. In some cases, their determination might be a difficult, if not an impossible, task.

4.1.2.2 Soft glassy model

Many soft materials, such as foams, dense emulsions, and textured morphologies of liquid crystals, are characterized by the presence of structural disorders on a mesoscopic scale (nanometers to microns) causing slow dynamical material evolution (they are classified as *soft glassy materials, SGM,* [19]). Their structural organization is thought to be metastable, i.e., out of the equilibrium. On the other hand, such materials exhibit rheological properties—therefore, models originating from the studies of the glass transition can be involved in predicting their rheological properties. The empirical criteria that define this class of materials are as follows [20]:

(a) They are very soft, and both elastic and loss moduli increase with the same weak power-law dependencies upon frequency.
(b) Hysteresivity (defined as a ratio of loss to elastic moduli) is frequency insensitive.

(c) At a given frequency, both elastic and loss moduli increase linearly with an increasing pre-stress.

(d) Under certain conditions, these materials display aging behavior.

The non-linear behavior of SGMs is often described using power-law equation between stress σ and strain ε, obtained using phenomenological approach:

$$\sigma = A + B \cdot \left(\frac{d\varepsilon}{dt} \right)^{\alpha}, \tag{4.17}$$

where A and B are constants corresponding to yield stress and consistency, respectively, α is the power present in the power-law relation with $\alpha < 1$ and $\alpha = 1$ describing the Hershel–Bulkeley [21] and the Bingham [22] fluid behavior, correspondingly.

The cytoskeleton of a cell is a system that integrates various molecules maintaining shape, integrity, and spatial organization of cells. Observing the similarity between the experimental data on cells and those reported for soft glassy materials, Fabry et al. hypothesized that the cellular cytoskeleton shares the same materials properties and behavior as SGM and thus soft glassy model can be applied to describe cell mechanics [23–25]. The conclusion has been made based on the measurements of various cells by means of magnetic twisting cytometry, where both elastic G′ and loss G″ moduli were determined. The measured mechanical properties correspond to the linear mechanical behavior of cytoskeleton embedded inside the cell and, therefore, both moduli are not describing the mechanical properties of cytoskeleton alone [26]. Based on the obtained results, Fabry et al. proposed an empirical equation that relates stress and strain (under conditions that $t \geq 0$):

$$\sigma(t) = \eta \cdot \delta(t) + \hat{G} \cdot \left(\frac{t}{\tau} \right)^{1-\alpha} \tag{4.18}$$

Equation (4.18) describes the stress response to a unit step strain imposed on cell at $t = 0$. In the equation, $\hat{G}$ is the ratio between stress and the unit strain measured at an arbitrary time τ, η is the Newtonian viscosity, and $\delta(t)$ is the Dirac delta function. The complex dynamic modulus G^* is given by the equation:

$$G^* = i \cdot \omega \cdot \eta + \hat{G} \cdot \left(\frac{i \cdot \omega}{\Phi} \right)^{x-1} \cdot \Gamma(x-2) \cdot [\cos \vartheta + i \cdot \sin \vartheta], \qquad (4.19)$$

where $\Phi = 1/\tau$, ω is the radian frequency, $(x - 1)$ is the power law exponent, $\theta = \tan(\pi \cdot (x - 1)/2)$ is the hysteresivity, Γ is the gamma function, and i is the imaginary unit ($i = \sqrt{(-1)}$). The equation is mathematically valid for $x < 2$ but the physical restriction set $x > 1$ (the relaxation function should not grow with time) what delivers the condition of $1 < x < 2$ (the material behavior undergoes transformation from the glass ($x = 1$) toward the liquid ($x = 2$) phases).

4.1.2.3 Tensegrity theory

The term "*tensegrity*" has been proposed by the architect Buckminster Fuller in 1961 [27]. It describes any kind of structures composed of tensional and compressional components being in equilibrium. These components exist in a state of constant pre-stress and, therefore, they mechanically stabilize each other (Fig. 4.4a).

The concept of tensegrity has also found many applications in cell biology. The single cell is modeled as a structure where tensional forces are generated in contractile acto-myosin filaments in the cytoskeleton [28]. Based on the importance of tensional pre-stress in maintaining cell shape and cellular stability, the cell can be modeled as two types of tensegral structures, i.e., the one being the cytoskeleton and the other one corresponding to cell nucleus (Fig. 4.4b).

These structures are counter-balanced by interconnected structural elements like microtubules or cross-linked actin filament bundles in the cell. The latter, fibrous structures resist compression together with the participation of other components, like for instance the extracellular matrix elements [29]. When there are no anchoring points, the shape of cell and nucleus is round and it is stabilized by internal tension. When cell is attached to a rigid substrate, the symmetric shape is not preserved any more.

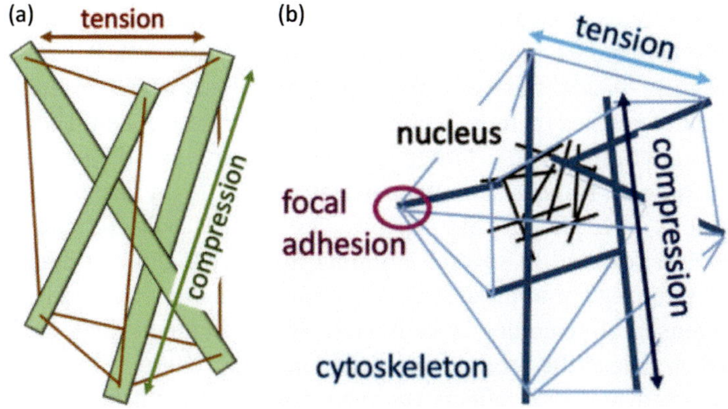

Figure 4.4 (a) Tensegral structure showing only tensional and compressional elements. (b) Single cell modeled with tensegrity approach. The model shows two types of interconnected tensegral structures with both tensional and compressional elements. Focal adhesions can be visualized in a tensegrity model as junction points of tensional elements with a compression one.

The tensegrity approach effectively predicts the mechanical behavior of various types of single molecules, cells, tissues, and whole organs. Various experiments confirm the applicability of tensegrity approach to describe the mechanical response of living cells [30–32]. The experimental verifications of the tensegrity model have revealed that mechanical forces applied at the macroscale are governed through discrete load-bearing cell adhesion sites that physically couple the contractile cytoskeleton to the extracellular matrix surrounding the cell. Integrins are the first molecules on the cell surface that sense mechanical signals and transmit it across the cell via specific molecular pathways. Since cells use the tensegrity approach to stabilize their shape, mechanical forces are transmitted through integrins to cytoskeletal elements present in the cytoplasm and nucleus. Importantly, due to faster wave propagation as compared to chemical diffusion, mechanical signals are transmitted much faster than signals of chemical origin. It has been shown that mechanical stress applied to integrin receptors resulted in almost immediate changes in the structure of the cytoplasm and nucleus and in activation of signaling events at distant sites.

The tensegrity approach has been widely applied to deliver a static cell model that does not consider any dynamical changes occurring inside the cell. In particular, cytoskeleton serving as a scaffold for a cell is highly dynamic structure and changes as a function of time. Moreover, the standard use of tensegrity only reflects static elastic properties of cells. Many experimental methods have shown that cell behaves as a viscoelastic material. This has resulted in several attempts to address the modification of the tensegrity approach. For example, Canadas et al. added to the tensegral structure of cells viscoelastic elements that can be used to describe creep phenomenon [33]. Here, the stress-supported network reflects the organization of actin filaments in the cytoskeleton (upon similar principles as tensegrity). The network has been used to analyze the effect of cell contractility considering elastic and frictional properties of the cell. Unfortunately, this approach describes only a limited number of features/phenomena observed in cells, presumably because of oversimplification in its assumptions. In particular, it fails to incorporate intermediate filaments that are known to participate strongly in cell mechanics.

4.1.2.4 Classification of material properties based on indentation

The main experimental techniques used to study material mechanical properties apply the external forces to induce/produce deformations [34]. Among them, nanoindentation-based methods, such as atomic force microscopy, cytoindenter, or cell/tissue poker, are widely applied to characterize cells and tissues mechanics. In all these methods, the relation between the load force and the resulting indentation is analyzed, since such dependence is characteristic for a given material. Thus, based on this relation, it is possible to classify the materials into three groups, namely elastic, plastic, and elasto-plastic ones (Fig. 4.5).

When a sufficient load is applied to a material, it will produce indentation, leading to changes in the material's shape. For *ideally elastic materials*, the shape returns do its initial state when the deforming force is removed. During the indentation experiment, both the load curve (i.e., force versus distance data collected during indentation) and unload one (recorded during indenter removal) overlap. When the applied external force is high enough to induce permanent deformations during indentation, the dependence between the load force and the resulting deformation are not reversible [35]. Such behavior is typical for *ideally plastic materials.* In reality, the mechanical response of various materials shows both contributions, i.e., elastic and plastic ones, manifested by the presence of hysteresis occurring in load–unload curves.

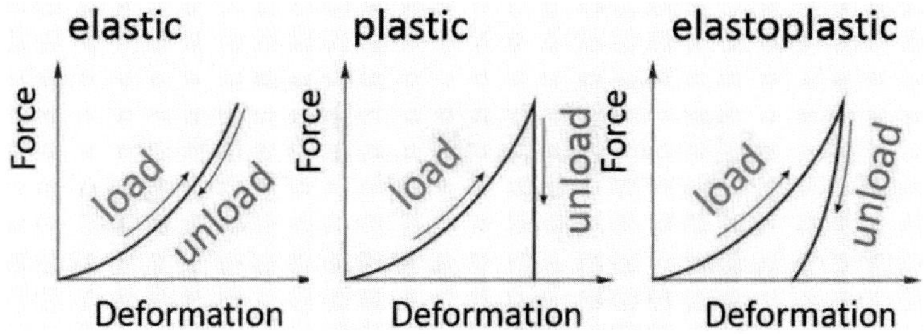

Figure 4.5 Force—deformation (indentation) relations characteristics for purely elastic, ideally plastic, and elasto-plastic materials.

4.2 Single-Cell Deformability Measurements

The mechanical properties of biological objects are closely related to their functions. The correlation between the cell ability to deform and some diseases has been known for a long time [36]. During last two decades, there have been a lot of research-reported differences between normal and pathologically altered cells and tissues, pointing out the correlation between mechanical properties and structural and morphological changes observed in various diseases [37, 38].

Large progress in biomechanics is a direct consequence of the development of techniques that enabled the deformability measurements in living cells at a single-cell level. As it has been recently pointed out, for example in [39, 40], currently available techniques can be classified into two groups: (i) methods that measure global properties of single cells (Fig. 4.6) and (ii) those that can probe single cell locally (Fig. 4.6).

Using techniques that deform a whole single cell, the overall mechanical properties can be determined. The most common methods are micropipette aspiration [41], microplate manipulation technique [42], microfluidic-based laminar flow [43], and optical tweezers [44].

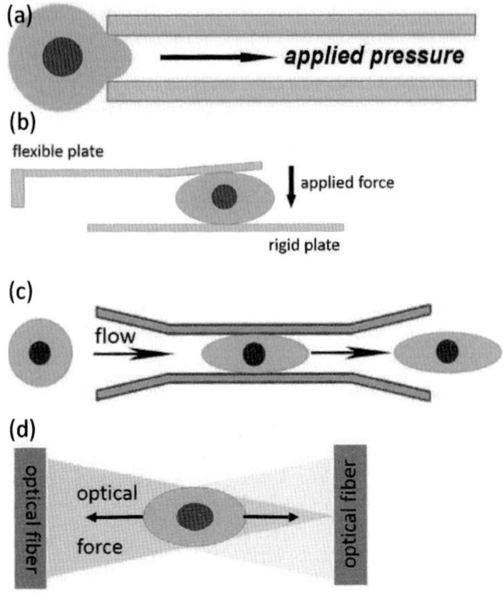

Figure 4.6 Common techniques used to measure deformability of a whole single cell: (a) micropipette aspiration, (b) microplate manipulation, (c) microfluidic flow systems, and (d) optical tweezers.

In micropipette aspiration, a single-cell deformation is generated by applying a small pressure applied into a glass pipette (Fig. 4.6a). A cell is soaked into the pipette with a diameter smaller than that of the cell. Changes in cell geometry, in particular its elongation, are related to the elasticity changes (expressed by Young's modulus) under assumption that there is no friction present between the cell membrane and the inner wall of the pipette. The magnitude of deformations is related to the applied suction pressure. In microplate manipulation, cells are deformed between two plates: the flexible and the fixed one. The load force can be determined by multiplying the deflection of a flexible plate by its spring constant (Fig. 4.6b). Knowing the relation between the load force and the measured deformation of a single cell, the elasticity modulus can be calculated. The deformability of single cells passing through narrow passages is the basis for most of the microfluidic systems. Here, the cellular deformation is caused by the laminar flow that is applied directly to cells. The strength of the flow is usually adjusted to allow a cell to pass along a narrow passage (Fig. 4.6c). In the fourth method, i.e., in optical tweezers, cells are deformed by forces induced by the momentum transfer from two counter-propagating laser beams to the surface of an individual suspended cell. (Fig. 4.6d).

The other group of techniques encompasses all methods that enable to measure mechanical properties locally. Here, the overall estimation of global deformability of single cells is derived from averaging of all data recorded at single locations on a cell surface. The three most popular methods are particle-tracking systems [45], magnetic twisting cytometry [46], and atomic force microscopy [38]. In the particle-tracking systems (Fig. 4.7a), the movement of particles (or natural small organelles like vesicles) specifically embedded inside the cell is traced and quantitatively expressed as a mean-square displacement. The motion of the particles can be related to mechanical properties of cell interior; however, thermal equilibrium must be assumed. In living cells, such assumption is not always true due to difficulties of distinguishing whether particle motion is driven thermally or, for example, by molecular motors. Tracing the movement of particles undergoing Brownian motion enables to study passive rheology-related processes. The use of externally applied optical or magnetic forces to move particles enable to study the active rheology-related processes.

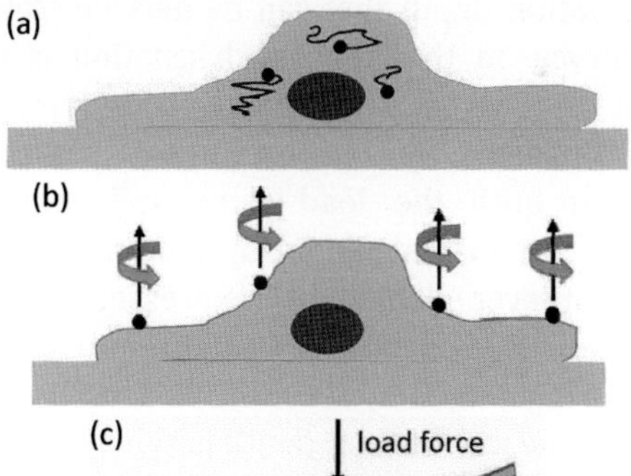

Figure 4.7 Three representative techniques that are widely applied to probe cell surface locally: (a) particle-tracking systems, (b) magnetic twisting cytometry, and (c) atomic force microscope.

An example of force-induced particle motion is magnetic twisting cytometry (Fig. 4.7b). Here, magnetic beads are coated with ligands that specifically bind to certain receptors (usually integrins) present on a cell surface. Then, externally applied magnetic force twists the magnetic beads. The resulting beads motion/rotation, inversely related to a cell stiffness, is then recorded and processed. The third technique used to probe cell surface is the atomic force microscopy (Fig. 4.7c). In this technique, the cell is indented by a probe mounted at the free end of a flexible cantilever. Based on the obtained force–indentation relations and the Hertz model of indentation, elastic modulus of a cell can be calculated.

4.2.1 Experimental Conditions for the AFM

Studies of single-cell mechanical properties require the AFM device working in force spectroscopy mode. In this mode, so called *force curves* are collected, as described in Chapter 3. To recall, a force curve denotes the dependence between the cantilever deflection and relative sample position that is next converted into force-distance curves. The estimation of the elastic properties of cells requires the knowledge on the relation between the load force and indentation depth that can be derived from the analysis of the force curves. In the AFM, such relation is not measured directly but it has to be determined. Therefore, the elasticity results are burden by uncertainties linked with the determination of both the load force and indentation depth values. The load force is calculated by multiplying the measured, uncalibrated cantilever deflection (expressed in volts) by the photodetector sensitivity (converting volts into nanometers) and by cantilever spring constant, resulting in force values expressed in nanonewtons. The indentation depth is calculated by subtracting the reference, calibration curve from the experimental data recorded on cells. As a consequence, the obtained elastic properties are relative. Thus, in order to reasonably use the AFM to study the mechanical properties of cells, always a calibration and a reference sample measurements must be carried out under the same experimental conditions.

In the experiment, the density of cells cultured on glass support should be low enough to preserve some area on glass support that is not covered by cells. This area will be further used for the calibration of photodetector sensitivity (PSD sensitivity). This measurement is particularly important because the PSD sensitivity is tightly dependent on the position of a laser beam at the end of free cantilever. Any changes in the laser position (also during, e.g., cells removal from the measurement's buffer) influence strongly the PSD value.

From the moment, when the calibrations of cantilever spring constant and photodetector sensitivity are completed, the acquisition of force curves on living cells may start (Fig. 4.8a). Having in mind that cells represent inhomogenous material, all measurements should be carried out in similar locations, i.e., either above cell nucleus or at the cell's periphery. The exemplary force curve recorded over a single fibroblast is presented in Fig. 4.8b (both approach and retract curves are shown). The reference curve recorded on hard, non-deformable glass surface is shown in Fig. 4.8c.

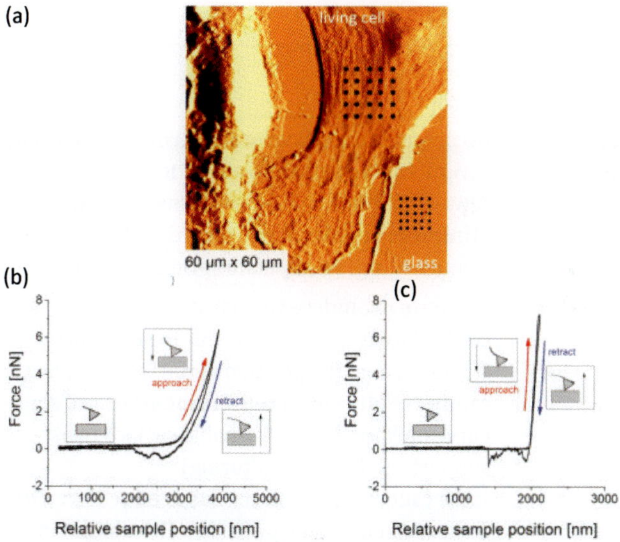

Figure 4.8 (a) The force curves acquisition. Cell is measured in a matrix of 5 × 5 indentations. Below, similar matrix of calibration curves, is collected on a glass substrate under the same experimental conditions. (b) Force curve obtained for a living cell. Both approach and retract parts are shown. (c) The calibration force curve acquired on a glass reference substrate.

In force curves shown in Fig. 4.8b,c, adhesive forces are observed during a cantilever withdrawing, regardless of surface type (i.e., either on glass or on cell). Table 4.1 summarizes the most important requirements, needed to be fulfilled during the AFM elasticity measurements, that are frequently time-consuming. Thus, obvious questions about statistics need always to be answered. It is rather clear that within one experimental day one should measure as many force curves or cells or samples as possible. Unfortunately, not always the collected set of data is big enough to be representative. Thus, if the results should answer the specific question about the local elasticity changes within a single cell, the focus should be placed on the acquisition of cell elasticity map with the largest possible number of curves. If the main aim is to compare changes induced within a population of cells, one can decrease number of force curves recorded on a single cell but, simultaneously, increase the number of studied single cells.

Table 4.1 Experimental requirements for the AFM elasticity measurements

Experimental requirements	Advice
Cantilever spring calibration	Do not use nominal values if during the experiment several cantilevers are planned to be used Always calibrate cantilever spring constants **Minimum requirement**: record cantilever resonance frequency and re-calculate spring constant
Photodetector sensitivity	Calibrate photodetector sensitivity under the same conditions as foreseen for cells measurement Do not change the position of the laser on a particular cantilever **Minimum requirement**: record single force curve on hard, non-deformable surface (glass or plastic support used for cell culturing) If the calibrated cantilever has to be removed from the measurement buffer, perform new calibration.
Cell cultures	Do not culture too many cells on a support—cells are secreting extracellular proteins that influences the quality of the calibration force curve (the photodetector sensitivity will not be determined reliably)

Experimental requirements	Advice
Changing samples	Always check photodetector sensitivity and cantilever resonant frequency to verify whether removing cantilever from the buffer did not destroy it
Changing cantilevers	While changing to new cantilever, always start with calibration
Statistics	Measure as many cells and samples as possible using particular cantilever. Most reliable comparisons can be obtained without changing the cantilever
Elasticity changes within a single cell	The number of measured cells can be reduced Acquire as dense elasticity maps as possible on a single cell **Minimum requirement:** a map of at least 8 × 8 points, the larger the better
Elasticity changes within cells' population	The number of force curves recorded for single cell can be reduced Acquire data for as many cells as possible

4.2.2 Criteria for Force Curve Selection

During automatic force curve collection operator does not have direct control over the device during the measurements—one can set only initial parameters that are maintained constant (depending on the stability of the AFM system used). Experimental results comprise therefore set of curves of variable quality that requires further processing in order to reject doubtful cases and select only valuable data. Figure 4.9 presents typical, a so-called "*good*" force curve.

The term "*good*" force curve denotes the curve where base line is represented as a straight, horizontal line (shadowed part in Fig. 4.9) and it is clearly distinguishable from the part of the curve corresponding to cell deformation (i.e., "deformed" part in Fig. 4.9). The ideal force curve is presented in Fig. 4.9a. In this case there is no adhesion observed during cantilever withdrawing. In practice, also force curves with small adhesion observed can also be also "good" (Fig. 4.9b). To record a full curve, the range of the Z scanner should provide the complete detachment of the cantilever from the cell surface. Depending on the cell type and cantilever used, the range will be different—the operator should assure that force curves are recorded using a maximum, reasonable range. This will facilitate both the determination of the contact point between the probing tip and the cell surface and the elasticity modulus determination. Note that due to liquid, salty conditions the real contact of the tip end with the cell surface may be difficult to be exactly estimated.

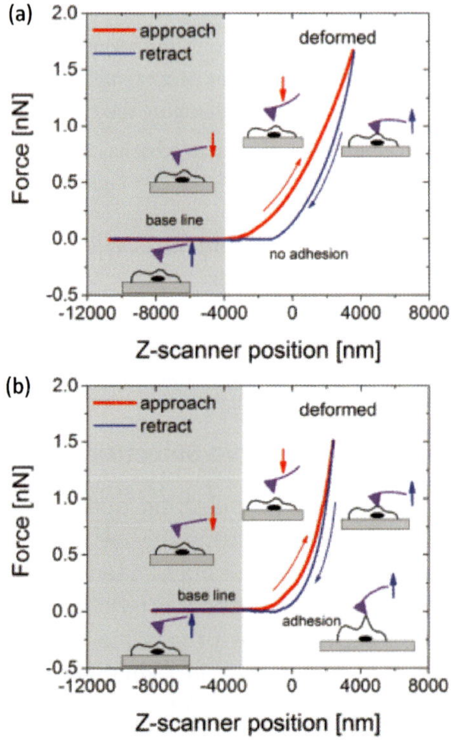

Figure 4.9 Force curves recorded on a single PZHPV-7 cell. Small pictograms show the behavior of the cantilever versus cell surface. Red curve represents cantilever approach, while the blue one retract. The horizontal base line corresponds to lack of interaction. Two examples are shown: (a) without adhesion, (b) with small adhesion.

The spectrum of possible force curves recorded during the indentation experiments on living cells is very broad (Fig. 4.10).

Therefore, the first step of data selection is based on the examination of the character of the recorded force curves. For example, the presence of small air bubbles on a cell surface (often present after a buffer exchange, in particular, on tissue surfaces) can be visible as a curve similar to the calibration one but characterized by the lower slope (Fig. 4.10a).

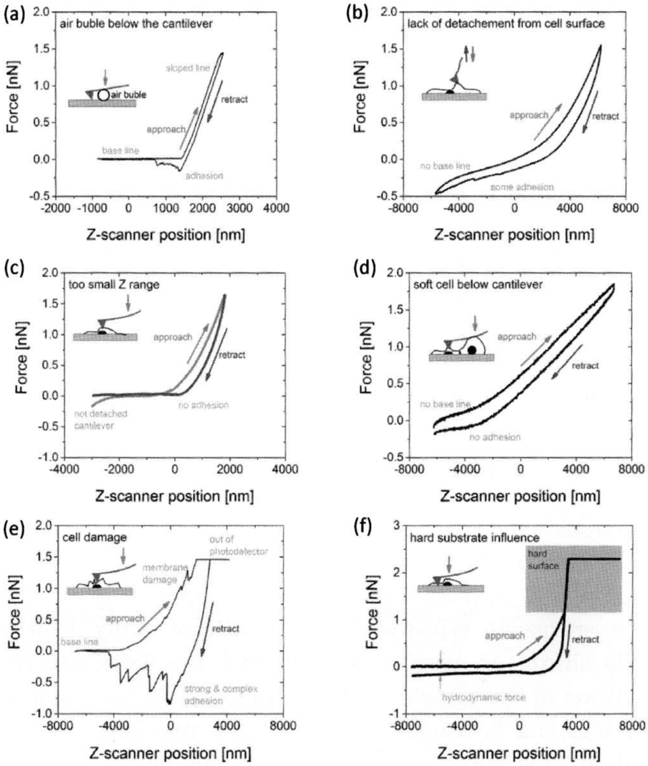

Figure 4.10 Examples of force curves recorded on single cells, showing typical unwanted features influencing the determination of cell elastic properties. Horizontal line observed at the maximum load (e, f) indicates that a reflected laser beam moved off the active area of the photodetector (observed frequently when full Z scanner range is used).

Depending on cell type (or more specifically on its adhesive properties), the range of the Z scanner should be adjusted to allow the full detachment of the probing cantilever from the cell surface. If the AFM probe is not fully detached, the base line (i.e., part of the force curve that should be represented by a straight, horizontal line) is sloped (Fig. 4.10b). Young's modulus calculated from such curve represents the effective modulus of the cell and some additional stiffness, originating from the membrane connection with the cantilever (note that, depending on the cell type, membrane tethers can reach few micrometer length).

The next example of "bad" data are cases when apparent full detachment is observed in the retraction part of the force curve but during recording of the next curve one can see that cantilever is still attached to cell membrane (Fig. 4.10c).

Sometimes, when the experiments are carried out on a cell monolayer, it can happen that an additional, second cell is placed below the cantilever (Fig. 4.10d).

Finally, setting a to high maximum load usually can lead either to cellular damage (ragged curves in Fig. 4.10e) or to the appearance of a two-slopped curve with the lower part corresponding to cell indentation followed by a straight line with slope slightly lower than the slope of the calibration curve, reflecting the spring constant of the used cantilever (Fig. 4.10f).

4.2.3 Force versus Indentation Curves

Force versus indentation curves are calculated by the subtraction from the measured data a reference deflection, measured on stiff and compliant surfaces (typically, only approach part of force curve is analyzed). When a stiff material (not easily deformable, typically silicon or glass) is investigated, the deflection reflects the position of the sample. In such case, force curve is a straight–sloped line and is usually employed as a reference line that is needed for the force calibration (red line in Fig. 4.11a). For compliant samples, like cells, cantilever deflections are much smaller and the resulting force curve has a non–linear character (blue line in Fig. 4.11b).

The difference between these two curves determines the deformation of the sample surface. For a given load force F_i, the indentation Δz_i is calculated as a difference between z_{xi} positions:

$$\Delta z_i = z_{2i} - z_{1i}, \tag{4.20}$$

where $z_i = F_i/k_{\text{cantilever}}$. As a result, the force versus indentation curve is determined (Fig. 4.11b). It describes the mechanical response to the applied load force and it is characteristic for each material.

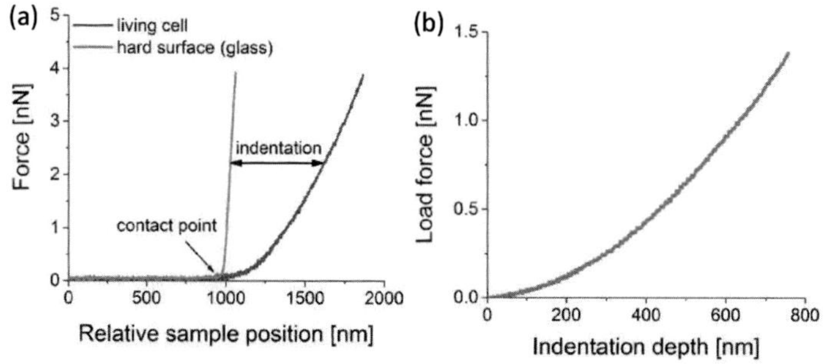

Figure 4.11 (a) The idea of the force–indentation dependence determination, (b) the resulting, exemplary force–indentation curve, calculated for a living keratinocyte.

4.2.4 Determination of Young's Modulus

The relative Young's modulus value, characterizing the cell stiffness, is usually evaluated in the frame of the Hertz contact mechanics [47]. It describes the indentation of two purely elastic spheres that can be extended into a case when a sphere indents infinitive thick, elastic half-space. The Hertz contact theory has been further developed [48], taking into account an infinitely stiff indenter with a chosen, approximate geometry of the AFM tip (i.e., spherical, paraboloidal, conical or flat-ended shapes) and a flat, deformable substrate. Usually, the AFM probe tip is a four-sided pyramid that is typically modeled either by cone or by paraboloid. This results in two equations delivering the relationship between the load force and the resulting indentation depth:

$$\text{Cone: } F(\delta) = \frac{2 \cdot \tan\alpha \cdot E'}{\pi} \cdot (\delta)^2 \tag{4.21}$$

$$\text{Paraboloid: } F(\delta) = \frac{4 \cdot \sqrt{R} \cdot E'}{3} \cdot (\delta)^{3/2}, \tag{4.22}$$

where F is the load force, δ is the indentation depth, α is the opening angle of the cone, and R is the radius of curvature of the AFM probing tip.

The E' is the reduced Young's modulus described by the relation

$$\frac{1}{E'} = \frac{(1-\mu_{\text{tip}}^2)}{E_{\text{tip}}} + \frac{(1-\mu_{\text{sample}}^2)}{E_{\text{sample}}}. \tag{4.23}$$

When $E_{sample} << E_{tip}$ (this is the case of living cells),

$$E' = \frac{E_{sample}}{(1-\mu^2_{sample})}, \tag{4.24}$$

where μ_{sample} is the Poisson ratio, representing compressibility of the sample material and ranging from 0 to 0.5. The Poisson ratio for cells is unknown and difficult to determine, therefore in all calculations its value needs to be assumed. Very often this value is simply set to 0.5, since cells can be treated as incompressible material.

In the literature [49], one can find also a less common equation describing the relation between force and indentation for a pyramidal shape of the AFM probe:

$$\text{Four sided pyramid: } F(\delta) = \frac{\tan\alpha \cdot E'}{\sqrt{2}} \cdot (\delta)^2 \tag{4.25}$$

The equation applies for symmetric pyramids.

During the analysis of the force–indentation curves, the fitted function is assumed to take a form of the power law $y = a \cdot x^b$, where the b value is set depending on the assumed shape of the indenting AFM tip. The resulting fit very often follows the quadratic function but this is not always the case. Sometimes, force indentations curves are better described by $y = a \cdot x^{1.5}$ (Fig. 4.12). To choose which model fits better, the goodness of the fit, χ^2 can be employed.

For example, the χ^2 values for the force–indentation curve obtained during the indentation of human bladder BC3726 cells were 0.0082 and 0.0614 for functions modeling the AFM probe as a paraboloid and cone, respectively. Following the procedure reported by Weisenhorn et al. [50], the tip shape approximation can be chosen from the values obtained during the fit of force versus indentation curve by the power law function: $y = a \cdot x^b$.

For the data presented in Fig. 4.12a, the fitted b value was in the range of 1.61 ± 0.17. Therefore, all recorded force curves were analyzed assuming that the AFM pyramidal tip can be modeled as a paraboloid (the parameter b was set to be 1.5).

Other cell type, prostate Du145, is better described by the function $y = a \cdot x^2$, corresponding to either a cone or pyramid model (Fig. 4.12b). Thus both equations can be applied; however, such calculations produce different modulus values, i.e., 1.77 kPa versus 1.48 kPa, respectively.

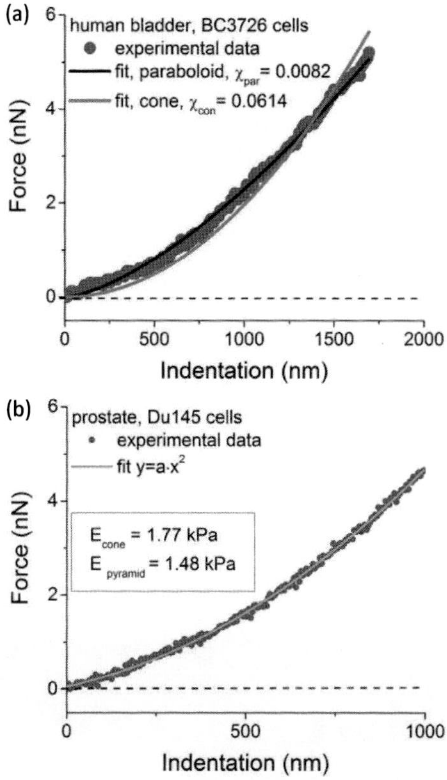

Figure 4.12 (a) The force–indentation curves recorded for living human bladder cells revealing the mechanical response of the studied cell. The response to deformation induced by four-sided pyramid better follows the equation describing the indentation using the paraboloidal indenter. (b) Young's modulus value of prostate Du145 cells calculated from the fit of $y = a \cdot x^2$ using equations for cone (4.21) and pyramid (4.23).

4.2.4.1 The final Young's modulus calculations

Independently of the theoretical model applied, the final Young's modulus is calculated taking into account all values determined for the whole subset of "good" force versus indentation curves recorded for single cells. There are two approaches. Either Young's modulus is estimated from all force curves measured for all cells or it is determined as a mean value of results calculated separately for every measured single cell. If the population of cells is homogenous, both approaches will deliver the same modulus values.

This is no longer true in the case of heterogeneous cell populations. As shown in Fig. 4.13, the final Young's modulus can vary twofold depending on the way how it was calculated.

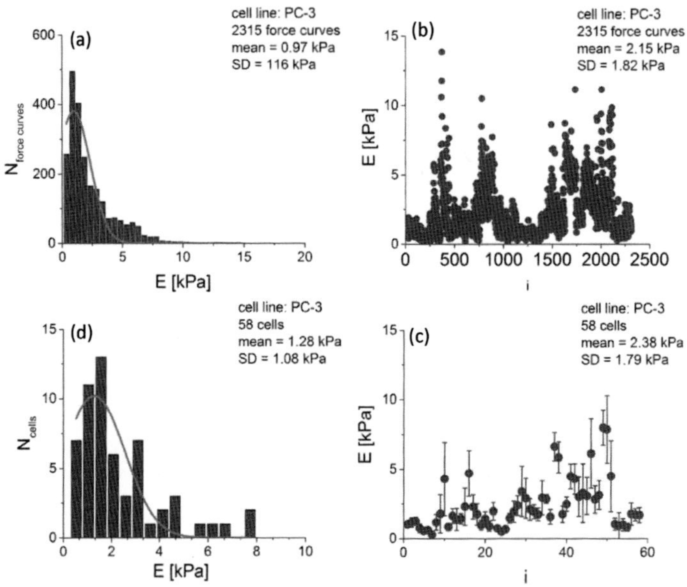

Figure 4.13 An exemplary analysis of the final Young's modulus, E, calculated for PC-3 prostate cells: (a) by fitting of the Gauss function to a histogram containing all moduli calculated for each force curve separately (total number of the recorded force curves is 2815). The center of the distribution denotes the most probable elastic modulus value and its width denotes a standard deviation (SD); (b) by calculating the mean value ± SD for moduli determined for each force curve separately; in this case, all modulus values are plotted as a function of consecutive force curve number (i from 1 to 2315); (c) by fitting the Gauss function to a histogram of Young's modulus obtained for each cell separately (N = 20 cells); (d) by calculating the mean value ± SD for moduli determined for each cell separately (the modulus is plot as a function of consecutive cell number, i from 1 to 20).

4.2.5 Depth-Sensing Analysis

In many biological applications, where the force spectroscopy mode is involved in the cell elasticity measurements, the commonly applied cantilevers have the spring constant in the range of 0.01–0.1 N/m and the typical loading force value does not exceed 30 nN [38]. In this range both glass and mica surfaces can be used as hard, non-deformable samples. When cells are probed, the resulting cantilever deflections are much smaller, indicating that the probing tip indents the cell. Depending on the magnitude of the indentation depth, distinct properties can be studied, revealing heterogenic structure of cell interior (Fig. 4.14).

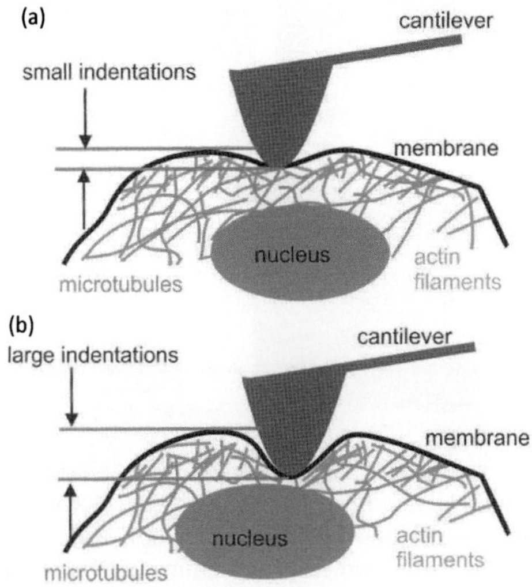

Figure 4.14 The idea behind the depth-sensing analysis that delivers the information on heterogeneity in the cell interior structure. (a) For small deformations the mechanical response of the cell is dominated by the actin cytoskeleton, while (b) in case of larger deformations, the overall stiffness of a whole cell is obtained.

As an example, the comparison between two cell types will be provided here [51]. Fibroblasts are the cells characterized by highly organized internal structure with well-differentiated both actin and microtubule cytoskeleton (Fig. 4.15a). Actin filaments are dispersed within the entire cell but they are mostly concentrated in the cortex layer beneath the cell membrane. They are organized into two groups: (i) stress fibers visible as a long and thick fibers and (ii) short actin filaments whose presence is barely detected under the fluorescent microscope. Microtubules extend from location close to cell nucleus toward membrane. The length of these cytoskeletal elements can reach even more than 100 microns. The incubation of fibroblasts with 5 μg/ml cytochalasin D leads to depolymerization of actin filaments and, as a consequence, to more homogenous spatial distribution of actin filaments (no stress fibers visible, Fig. 4.15b).

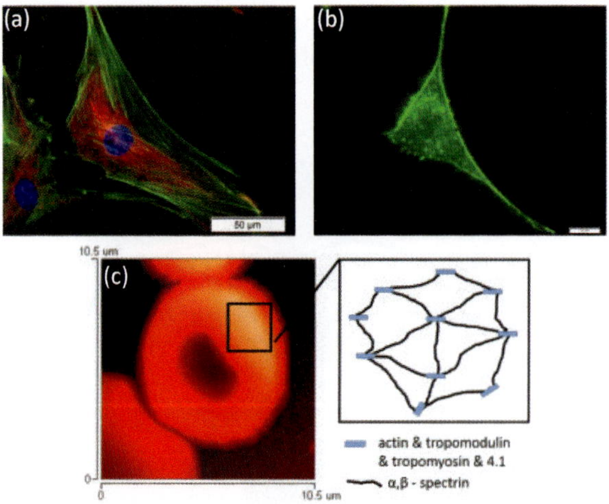

Figure 4.15 The structure of (a) a single fibroblasts visualized using fluorescence microscopy (green—F-actin stained with phalloidin labeled with Alexa Fluor 488, red—microtubules labeled using two-step procedure involving antibody against β-tubulin followed by the secondary antibody labeled with Alexa Fluor-555, and blue—cell nucleus stained with Hoechst dye); (b) Actin cytoskeleton of living fibroblasts after incubation with 5 μg/ml cytochalasin D for 10 minutes at room temperature; (c) Surface topography of a single erythrocyte recorded using atomic force microscopy, accompanied by a scheme of its cytoskeleton being a network in which edges are spectrin filaments interconnected through various proteins (actin, protein 4.1, tropomyosin, etc.).

Erythrocytes are more homogenous as compared to fibroblasts (Fig. 4.15c). There is no nucleus and the cytoskeleton is composed of a filamentous meshwork of proteins that forms a membrane skeleton along the entire cytoplasmic surface of the membrane.

The cytoskeletal filaments are composed of spectrin, forming long, flexible heterodimers through the lateral association of α and β spectrins linked to junctional complexes composed of F-actin, protein 4.1, and actin-binding proteins (like tropomyosin or tropomodulin).

By the depth-dependent analysis of mechanical properties, the effect of cytoskeletal structure and organization on cellular elasticity can be explored. Figure 4.16 present the dependence of the average Young's modulus calculated as a function of indentation depth for fibroblasts (red circles), erythrocytes (blue squares), and fibroblasts treated with cytochalasin D (dark cyan stars), respectively.

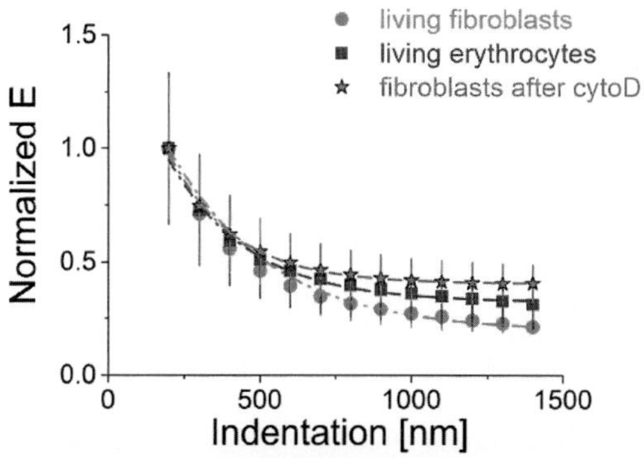

Figure 4.16 Elastic modulus plotted as a function of indentation depth for fibroblasts, erythrocytes, and fibroblasts treated with cytochalasin D agent (cytoD, final concentration of 5 μg/ml) leading to depolymerization of actin filaments). Modulus values correspond to indentation depths increasing by 200 nm and are presented as a mean value ± standard deviation, i.e., 4.85 ± 2.03 kPa, 1.38 ± 0.12 kPa, and 2.48 ± 1.49 kPa, for fibroblasts, erythrocytes and fibroblasts treated with cytochalasin D, respectively.

These relationships are fitted with the simple exponential decay function ($y = y_0 + A \cdot \exp(-ind/s)$ that enables to estimate the decay rate (expressed in pm^{-1}) used to quantify the degree of cytoskeleton organization. The obtained values are 30.0 ± 0.3 pm^{-1}, 38.2 ± 0.3 pm^{-1}, 46.7 ± 0.2 pm^{-1} for fibroblasts, erythrocytes, and fibroblasts treated with cytochalasin D, respectively. As mentioned previously, the cytoskeleton of fibroblasts is highly heterogeneous while of erythrocytes is highly homogenous. The decay rate increases with homogeneity of the cytoskeleton organization.

Large values of standard deviation, observed for smaller indentation depths, are the manifestation of Young's modulus distributions as it can be seen in Fig. 4.17 showing histograms for two limit values of the indentation depths, namely, 200 nm and 1400 nm. The widest distributions of elastic modulus are observed for small indentation depths (for instance, for 200 nm, Fig. 4.17a,c). Distributions become narrower with the increase of the indentation depth (for example, for 1400 nm, Fig. 4.17b,d). The depth-sensing analysis suggests that the very local AFM measurements of cell's mechanical properties, carried out at high lateral and force resolution, detect small/minute changes in the organization of cytoskeletal network that can be explained by its filamentous nature.

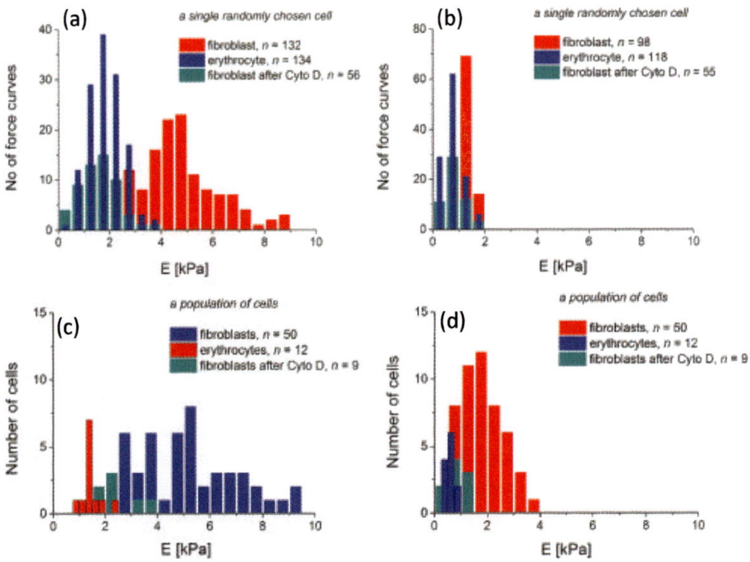

Figure 4.17 Histograms of the relative values of Young's modulus obtained for living fibroblasts (red columns), fibroblasts treated with cytochalasin D (dark cyan columns), and erythrocytes (red columns). Typical modulus distributions of a single cell obtained for indentation depths: 200 nm (a) and 1400 nm (b). Elastic modulus distributions for cell populations obtained analogously (c, d for 200 and 1400 nm, respectively). Force versus indentation curves were fitted assuming that the AFM tip can be represented as a cone. Bin size = 0.5 kPa, n denotes either the number of force curves or cells taken for the analysis. Reprinted with permission from [51].

The observed heterogeneity in Young's modulus observed at small indentations depths can be explained by several phenomena, including non-homogenous cytoskeleton density, removal of a limited number of filaments composing the network or even disruption of cytoskeletal filaments induced by AFM indentations. The narrowing of histograms as indentation depths increases stems from the fact that for larger indentations the information on cell mechanical properties is gathered from a larger volume. In this limit, elastic modulus should reach constant level at very large indentations, above 1–2 microns. Distributions presented in Fig. 4.17 show also cell-type relationship. Fibroblasts, being the cells with highly differentiated cytoskeleton and heterogeneous organization of filaments, reveal wider distribution of the modulus values. The modulus distribution for erythrocytes is much narrower. The incubation of fibroblasts with cytochalasin D, an agent depolymerizing actin filaments, manifests in the narrowest modulus histogram. This relation is observed both when a single cell or a population of cells are considered (Fig. 4.17).

To quantify the observed changes, a comparative parameter R has been introduced by Pogoda et al. [51]. It is defined as a ratio between the modulus values calculated for two limits of indentation depths: 1200 nm and 200 nm. In the original paper, its value was determined for randomly chosen single cells. Figure 4.18 shows parameter R distribution calculated for all studied cells (still being defined as a ratio E_{200} and E_{1200}).

In the analysis, histograms corresponding to each studied cell type were created and to each one the Gauss function was fitted in order to get the center position of the distribution denoting the mean value. The comparison of the *mean values ± standard deviations* for all analyzed cell types is shown in Fig. 4.18b. Depending on the cell type, erythrocytes and fibroblasts, two distinct values of R parameter are observed as it was confirmed by the Student t-test ($p < 0.001$, at the 0.05 level). The treatment of fibroblasts with the cytochalasin D shows slight increase of the R value, accompanied by a low standard deviation value indicating high degree of homogeneity in elastic properties of cells. The Student t-test showed non-significant difference between non-treated and cytochalasin D treated fibroblasts which became statistically significant when compared with erythrocytes (at the 0.05 level).

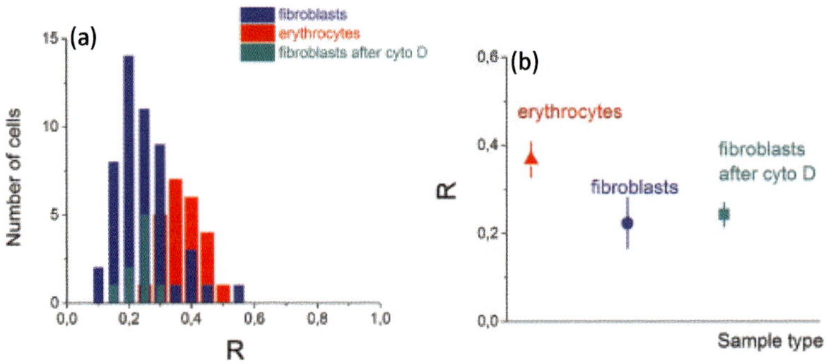

Figure 4.18 (a) Exemplary distributions of a parameter R (defined as a ratio E_{1200} and E_{200}) obtained for erythrocytes, fibroblasts, and fibroblasts treated with cytochalasin D. (b) Mean values of R parameter determined for a given cell sample types. Errors are standard deviations.

The largest difference between elastic modulus values for small and large indentation depths manifests in low R value. More homogenous internal cellular structure shows much smaller variations in the depth-dependent modulus values and thus R takes larger values, with the limit at 1 (when modulus is independent on the indentation depth). This case is observed for fibroblasts and erythrocytes. The treatment of fibroblasts with cytochalasin D increases R value, but still the internal structure inside such cell is far from being homogenous.

4.2.6 Stiffness Tomography

The analysis of mechanical properties as a function of indentation depth is also the basis for *stiffness tomography,* introduced as a new imaging AFM functionality in 2009 by Roduit et al. [52]. Typically, as in the AFM force spectroscopy mode, force curves are recorded in relation to the position on the investigated surface (Fig. 4.19a).

The indentation is carried out up to a given value of the load force. Depending on the cell structure probed, if, e.g., stiffer inclusions (like actin filaments) are probed beneath cell membrane the resulting force–indentation relation can manifest as a steeper curve (corresponding to larger Young's modulus value). Under the assumption that the presence of harder inclusions changes the shape of the force curve significantly, the obtained force–

indentation relations can be segmented into slices [51–53]. To each slice the Hertz model is applied and the corresponding Young's modulus value is calculated. The obtained data are plotted as a function of indentation depth (Fig. 4.19b). The stiffness tomography approach has been initially applied to study the 3D variability in distribution of Young's modulus in living neurons [52]. The results showed a presence of hard structures, attributed to cortical actin cytoskeleton, whereas softer regions corresponded to surrounding cell membrane. The further applications demonstrated feasibility of the stiffness tomography approach to study the mechanical properties of bacterial membranes, evidencing the presence of stiffer regions lying beneath membrane. The possible explanation involved the hypothesis of various structures accumulations in certain region in the cytoplasm [53].

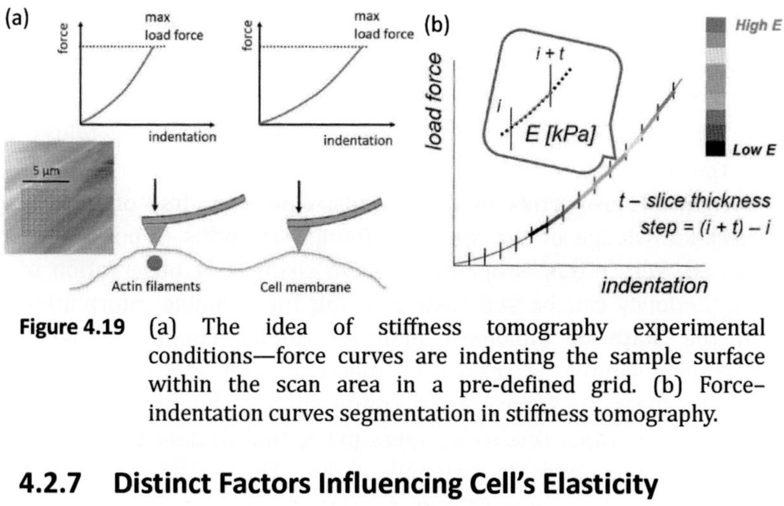

Figure 4.19 (a) The idea of stiffness tomography experimental conditions—force curves are indenting the sample surface within the scan area in a pre-defined grid. (b) Force–indentation curves segmentation in stiffness tomography.

4.2.7 Distinct Factors Influencing Cell's Elasticity

The use of the AFM to determine Young's modulus has several limitations that, in fact, made this method to be extremely difficult in the determination of the modulus *absolute* value. The most important factors influencing cells elasticity can be grouped into several classes related to

(a) the applied calibration methodology linked with the uncertainties in the determination of cantilever spring constant and photodetector sensitivity;

(b) the variability of cell-related factors such as culture conditions (buffer composition), density of cells, (if they are touching each other or not), confluence on a substrate, the number of passages, the day of measurement after the passage, etc.;
(c) the variability of the experimental conditions provided by the AFM device, such as load speed, place of poking, number of force curves recorded at one place, location of the measurements on cell surface, the presence of the stiff substrate below the investigated cells, etc.;
(d) the applied contact mechanics model used to describe cellular elastic properties (typically, the Hertz contact mechanics with Sneddon's modifications and cone approximation, etc.);
(e) the way of data analysis, especially the determination of the point of contact between the indenting AFM tip and cell surface, range of indentation depth or load force;
(f) the physical and chemical substrate properties influencing cell behavior.

All these sources of errors may raise doubts in usefulness of the AFM technique for measurements and analysis of the mechanical properties of living cells. However, first of all, the exact knowledge of the absolute Young's modulus is not always needed. Very often, either the relative changes or observation of the tendency can be sufficient to bring the valuable information on the state of biological material (under condition that a reference sample is measured together with the investigated ones in the same experimental conditions). It should be stressed here that multiple research papers prove that despite the various uncertainty sources mentioned above, the relation between healthy and cancerous cells is preserved. This has been shown for various cancers, including, for example, bladder [38], breast [54, 55], prostate [54, 56], ovarian [57], and thyroid [58] cancers.

4.2.7.1 Calibration-based discrepancy

The essential step in the determination of Young's modulus is the appropriate calibration of the cantilever spring constant and the photodetector sensitivity. The calibrated load force F is used in the fitted equation:

$$F(\text{Signal}\times\text{invPSD}\cdot k_{\text{cant}}) = a\cdot(\text{ind})^{b}, \tag{4.26}$$

where a is proportional to Young's modulus value. Thus, the errors introduced by discrepancies in both cantilever spring constant and photodetector sensitivity influence strongly the accuracy of the elastic modulus determination. The maximum error can be estimated by

$$\Delta a \propto \Delta E = \sqrt{\left|\frac{\partial a}{\partial k_{\text{cant}}}\right|^{2}\cdot\left|\Delta k_{\text{cant}}\right|^{2} + \left|\frac{\partial a}{\partial \text{invPSD}}\right|^{2}\cdot\left|\Delta\text{invPSD}\right|^{2}} \tag{4.27}$$

As an example estimation, for particular, typical case, uncertainties in spring constant introduce roughly 10% of total error, while the photodetector sensitivity part adds additional 14%.

4.2.7.2 Variability stemming from cell-related factors

Cells are highly dynamic in their properties. Therefore, usually, discrepancy arising from the applied calibration methodology can be less significant as compared to variations of Young's modulus stemming from cell-related factors, such as culture conditions (culture medium composition), density of cells, confluence on a substrate, the number of passages, the day of measurement after the passage, etc. Despite extensive research showing mechanical differences between normal and pathological cells, there is a little information devoted to systematic studies showing the effect of cell-related factor on single-cell biomechanics. Some examples are mentioned below and summarized in Fig. 4.20.

The influence of medium composition on mechanical properties of non-malignant (MCF10A) and metastatic (MDA-MB-231) breast cancer cells has been reported in the work of Nikkhah et al. [59]. Five different compositions were investigated, i.e., medium M1—RPMI supplemented with 10% FBS; M2—RPMI supplemented with 5% FBS; M3—RPMI supplemented with 5% FBS and 20 ng/ml epidermal growth factor; M4—DMEM:F12 supplemented with 5% FBS; and M5—DMEM:F12 supplemented with 5% horse serum, 20 ng/ml epidermal growth factor, 100 ng/ml cholera toxin, 0.01 mg/ml insulin and 500 ng/ml hydrocortisone (Fig. 4.22, *prepared using the data published in* [59]).

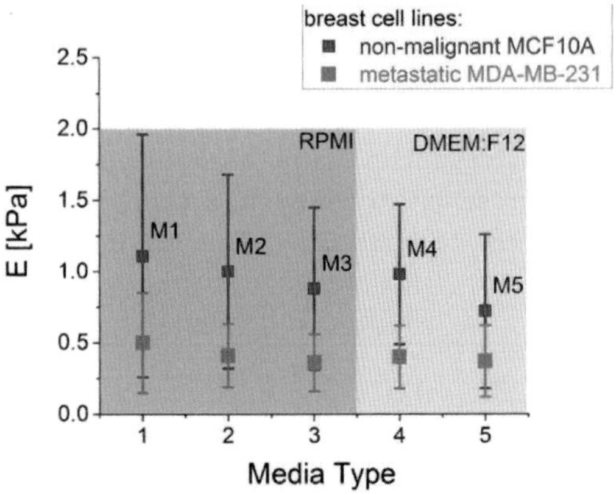

Figure 4.20 The elastic properties of breast cancer cells, determined for various medium composition: M1—RPMI supplemented with 10% FBS; M2—RPMI supplemented with 5% FBS; M3—RPMI supplemented with 5% FBS and 20 ng/ml EGF; M4—DMEM:F12 supplemented with 5% FBS; M5—DMEM:F12 supplemented with 5% HS, 20 ng/ml EGF, 100 ng/ml CT, 0.01 mg/ml INS and 500 ng/ml HC (FBS—fetal bovine serum; EGF—epidermal growth factor, HS—horse serum, CT—cholera toxin, INS—insulin, and HC—hydrocortisone).

The effect of medium composition on single-cell elastic properties is more dominant for non-malignant MCF10A breast cell lines as compared to the results obtained for metastatic MDA-MB-231 cells. Elastic modulus of non-malignant breast cells changes from 1.11 ± 0.85 kPa to 0.72 ± 0.54 kPa (decrease of about 35%). In case of metastatic MDA-MB-231 cells the modulus changes from 0.50 ± 0.35 kPa to 0.37 ± 0.25 kPa (~26%). These results strongly suggest the use of similar medium composition in studies aiming at the comparison of properties of various cell lines (or consider the effect of medium composition if it is not possible to use the same culture medium or use reference cell lines). Simultaneously, such studies could be accompanied by fluorescent images of actin cytoskeleton to bring conclusion whether culture medium composition changes the actin filament organization that is manifested in alterations of mechanical properties of single cells.

Some research demonstrates also the effect of cellular microenvironment on mechanical properties of single cells (Fig. 4.21).

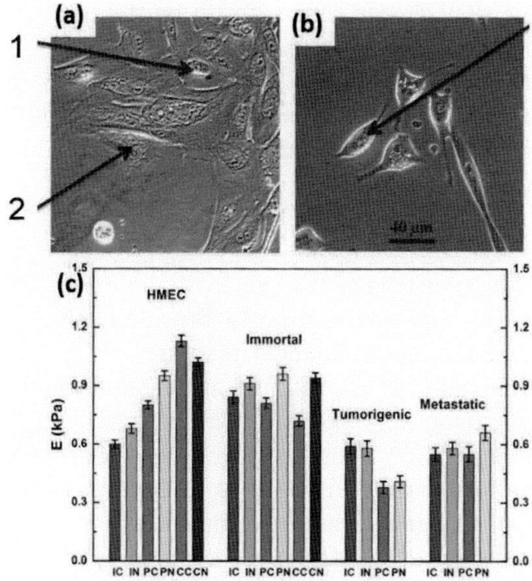

Figure 4.21 Bright-field images of cells in culture: (a) A small colony or monolayer. (b) Several individual cells (the arrows indicate a cell placed inside (1), and at the periphery (2) of a monolayer of cells, as well as an isolated cell (3)). (c) Elastic modulus (*mean ± standard error of the mean*), determined for normal (HMEC), immortal, tumorigenic, and metastatic cells measured in various conditions: I—isolated single cell; P—cell located at the periphery of a monolayer; C—cell measured inside of a monolayer (close to its center) and distinct places within the cell: C—cytoplasm; N—nucleus (*re-printed from [60]*).

For example, a recent study by Guo et al. reported the effect of neighboring cells on elastic properties of normal (human mammary epithelial cells, HMEC cells), immortal (derivatives of HMECs that were transfected with hTERT), tumorigenic (HMLER cells—HMECs cells taken at 23 population doubling time) and metastatic (MDA-MB-231 breast cancer) cells [60]. Depending on the place of measurement within a single cell and the density of cells, distinct elastic moduli are observed. Normal (HMEC) and immortal cells are stiffer than both tumorigenic and metastatic cells. They show larger Young's modulus values, ranging from 0.60 ± 0.05 to 1.13 ± 0.06 kPa, while the elastic modulus for tumorigenic and metastatic cells varies from 0.38 ± 0.03 to 0.66 ±

0.03 kPa. The tumorigenic cells seem to be the most deformable cells among the four studied types of breast cells (E = 0.49 ± 0.11 kPa versus E = 0.58 ± 0.05 kPa for tumorigenic and metastatic breast cells, respectively). Noteworthy, there is a small difference between modulus determined from areas above the nucleus and the cytoplasm (not exceeding 15%). Among four studied cell lines, only normal cells (HMEC) show a strong dependence on the cellular microenvironment. The largest Young's modulus is observed for cells located at the center of a monolayer where the neighboring cells influences strongly the mechanical properties (the elastic modulus almost doubled its value).

Most of animal cells are cultured at temperature of 37°C that resembles physiological conditions. On the other hand, majority of scientific papers report the elasticity measurements carried out at room temperature that varies, usually, between 20 and 22°C. Few reports show rather inconsistent effect of temperature on elastic properties of living cells. For example, NIH3T3 fibroblasts seemed to be unaffected within the temperature range 31–37°C, whereas the increase to 43°C causes sudden drop of modulus value. The 7-4 cells behaved differently. For them, a modulus maximum at 37°C was observed [61].

Another cell-related parameter influencing cell mechanical properties is the number of passages carried out before the measurements. In general, for immortalized cell lines, no differences are reported (see, e.g., [61]). However, for such cells, the cytoskeleton organization was probably preserved from one passage to another. One could expect some variations for normal and primary cell lines, whose doubling time is limited. In such case, the alterations in elastic properties for large number of passages are almost directly related to changes in actin cytoskeleton structure and cells state. Close to the maximum number of possible passages, cells are not dividing and their morphology is clearly distinguishable as compared to cells after first few passages.

4.2.7.3 The influence of the AFM experimental conditions

The determination of the elastic modulus can be influenced by the way how the AFM force spectroscopy is carried out. Thus, the knowledge how to setup the AFM experiment is vital to obtain reliable data. This includes choice of a size of the grid within which force curves are recorded, number of force curves recorded at one position, where to measure cellular deformability, load speed that is related to loading rate, maximum force load, etc.

The demand of high statistics requires a prolonged poking of every single cell, which can lead either to a damage of cell membrane or to remodeling of actin cytoskeleton (Fig. 4.22).

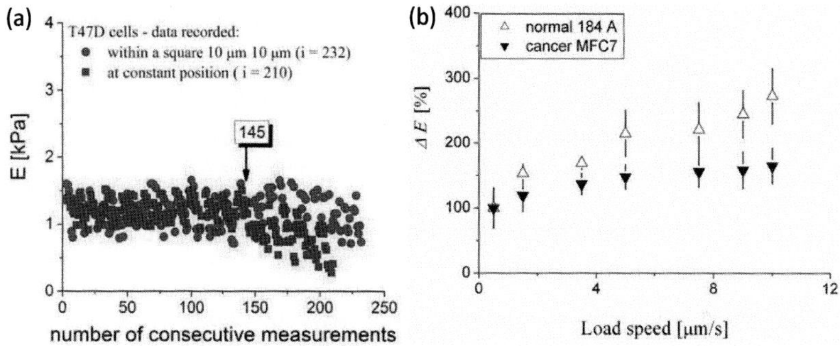

Figure 4.22 (a) The effect of prolonged poking at the single location on a single cell showing a drop of the elastic modulus value at certain number of a recorded single force curves. (b) The plot showing Young's modulus dependence on the load speed. Reprinted with permission from [55].

Both effects are undesirable and should be avoided since they can lead to changes in the elastic modulus. A convenient way to verify whether there is no influence of the prolonged cell poking in the obtained date is to plot Young's modulus as a function of the number of consecutive measurements (which is also related to the time elapsed during measurements of consecutive cells). Figure 4.22a presents such data obtained for TD47 breast cells [55]. Purple squares show the modulus calculated from force curves recorded during poking at predefined, constant position. The measured moduli, randomly distributed around the mean value of 1.20 ± 0.28 kPa start to decrease, revealing a breakdown at the curve no. 145 and indicating sudden change in the elastic modulus and thereby strong alterations in actin cytoskeleton. More stable data can be obtained by setting a square area, e.g., 10 μm × 10 μm, within which force curves are acquired (blue dots in Fig. 4.22a). Here, the prolonged poking does not induce/generate such clearly visible remodeling of actin cytoskeleton as compared to poking at single location.

Young's modulus determined by AFM is not a constant and absolute value—it is dependent on the loading rate [54, 55, 62]. In the AFM the loading rate is indirectly introduced by the load speed (the speed of cantilever approach, describing how fast a living cell is indented). As one can see from Fig. 4.22b, the effect of the indenting speed (i.e., loading rate) is cell-dependent. Normal breast cells (184 A) are more sensitive to load rate than cancerous MFC7 cells. Similar results have been already reported for another breast cancer pair, i.e., benign MCF10A and cancerous MCF7. Also in this case, MCF7 cancer cells were more deformable (lower modulus) and changed their elastic properties less pronouncedly as compared to benign cells [54].

4.2.7.4 Discrepancies stemming from the Hertz contact mechanics theory

The Hertz contact mechanics [47, 48] is widely applied in quantification of the AFM data despite the required assumptions that are only partially fulfilled:

(a) A cell is treated as a isotropic and purely elastic material;
(b) A cell can be approximated by an infinitely thick half space.
(c) There are no adhesion forces within the contact area between the probing tip and the cell surface.

These assumptions simplify Young's modulus determination but make it difficult to obtain the absolute modulus value and, simultaneously, introduce additional ambiguity in the estimated elastic properties. Some consequences of this issue are discussed below.

The theoretical model used for Young's modulus determination assumes a particular shape of the indenter. The shape of a typical AFM tip is a four-sided pyramid with a height varying from 2.5 μm to even 15 μm and a distinct tip apex ranging from 2 nm to 50 nm. The most popular approximations of the AFM tip shape relate the load force F and the indentation depth z as follows:

(1) $F \sim z^2$ (cone and pyramid)
(2) $F \sim z^{3/2}$ (paraboloid)

Due to heterogenic nature of cells, experimental data do not always follow the chosen theoretical model. In such case, one must bear in mind that imperfections in the fit are unavoidable and not much can be done about uncertainty in the final results.

Figure 4.23 shows fits obtained for indentation depths of 200 nm, 800 nm, and 1400 nm, carried out using two most common AFM tip geometries: paraboloid and cone.

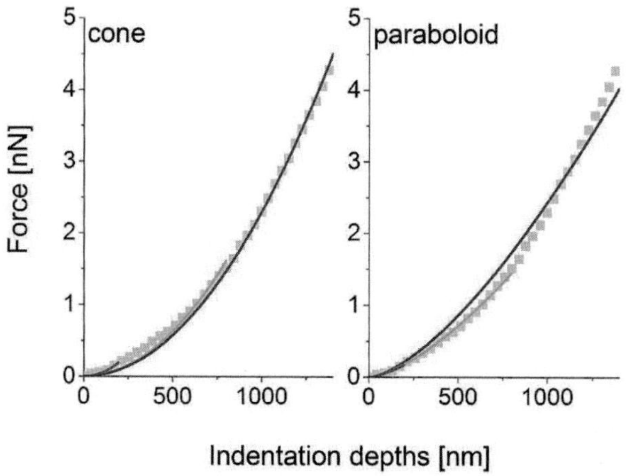

Figure 4.23 The function used to fit the experimental data is $y = a{\cdot}x^b$ where b denotes the assumed shape of the AFM tip (2 for cone and 3/2 for paraboloid) and a is the coefficient proportional to Young's modulus. Violet, orange and brown lines correspond to the fits performed for the indentations of 200, 800 and 1400 nm, respectively while gray points denote the experimental data. Reprinted with permission from [51].

Most of AFM experimental data show a kind of relationship contained in the area between two model shapes, i.e., cone and paraboloid, where sometimes the former sometimes the latter assumption fits better. Usually, the form of the fitted function is $y = a{\cdot}x^b$ where b is usually set to be either 2 either 3/2. To estimate more precisely how the performed fit affects the determination of Young's modulus, the following method can be applied. The experimental data may be represented by the following relation:

$$y_n = c \cdot x_n^{|b+\varepsilon|}, \tag{4.28}$$

where n is the number of point in the force–indentation curve, y_n is the force value corresponding to the indentation of x_n, c is the fitted parameter that is proportional to elastic modulus value, b includes the information on the shape of the AFM tip, and ε denotes the low value expressing how much the fitted line deviates from the theoretical model. By comparing the fitted function (Eq. 4.28) with the theoretical model $y = a \cdot x^b$, one can estimate that

$$\frac{c}{a} = x_n^{|\varepsilon|}. \tag{4.29}$$

For ε close to zero the fitted c parameter approximates better the coefficient a proportional to Young's modulus.

Geometrical properties of the AFM cantilever also influence modulus values estimation. The theoretical relation presented in Fig. 4.24 illustrates Young's modulus relation on load force calculated for three types of indenting probes: pyramidal, flat-ended, and spherical ones [61].

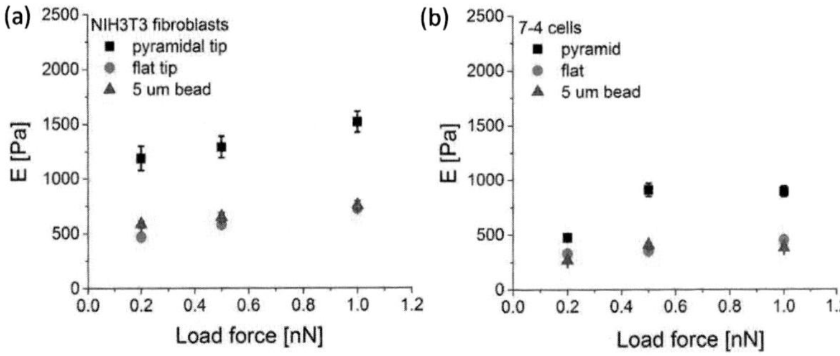

Figure 4.24 Elastic modulus variations as a function of load force calculated for two types of living cells: (a) NIH3T3 fibroblasts and (b) 7-4 (Ha-Ras^{V12} oncogene transformed) cells. Images prepared using data from [61].

From the presented comparison, one can clearly see that the use of pyramidal tip delivers larger elastic modulus values while values obtained for flat-ended and spherical (a bead with diameter of 5 μm) indenters are lower for a given load force. The increase of the indenting force manifests in gradual modulus increase, which is better visible for NIH3T3 fibroblasts than for 7-4 cells [61].

4.2.7.5 The contact point determination and data analysis

As it was mentioned earlier, in the AFM, the force versus indentation curves are calculated from the subtraction of calibration curve (recorded on hard surface) from a curve recorded on a cell) and. This approach can be strongly burdened by the choice of the contact point position, as is shown in Fig. 4.25.

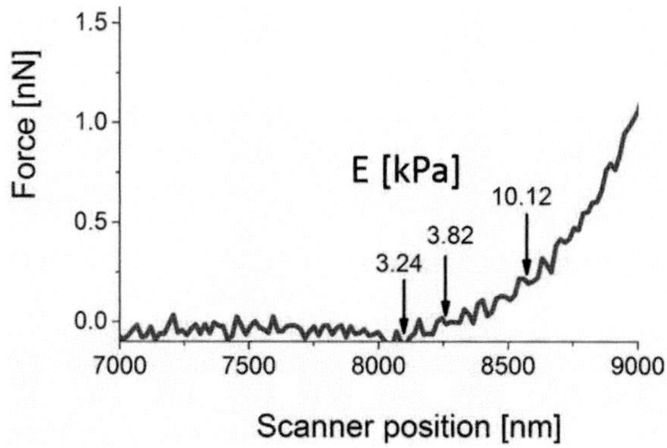

Figure 4.25 Elastic modulus variations as a function of the assumed position of contact point.

In the above, exemplary data, in the approach part of the force curve, three distinct locations of the contact point are marked: at 8086, 8279, and 8560 nm. The corresponding elastic moduli were 3.24, 3.82, and 10.12 kPa. For two first contact points, chosen within a small region of a horizontal baseline, the modulus alterations fall within 15–20%. Larger modulus variations are noted for the third contact point marked on the slopped part of the force curve—more than 250% increase is observed.

The example showing the data analysis evolution can be nicely demonstrated for two human bladder cells, i.e., HCV29 (non-malignant cell cancer of ureter and T24 (transitional cell carcinoma) cells. These cells were measured for the first time using AFM in 1999 [38], then in 2012 [66]. The paper from 1999 belongs to a group of early papers where details of cells' mechanical properties measurements were not deeply understood. For the low load force (up to 2 nN) the data were analyzed using approximation of the probing AFM tip by a paraboloid because the mechanical response suggested this assumption (the simple criterion was applied by checking chi square value—its smaller value indicated better fit). Young's modulus was calculated for two indentations: 600 nm for HCV29 cells and 1600 nm for T24 cells. In the paper from 2012, the elasticity modulus was calculated for the indentation depth of 500 nm for both cell lines. In the oldest study, the load force value was not considered at all.

The elastic modulus for HCV29 cells obtained in 1999 was 9.7 ± 3.6 kPa versus 2012 measurements of 3.19 ± 0.27 kPa. For T24 cells, older measurements delivered 0.8 ±0.4 kPa, while newer study value was 0.77 ± 0.25 kPa. This difference not only is a consequence of improved data analysis but also stems from various experimental conditions (cf. Table 4.2 in Section 4.2.7.7).

4.2.7.6 Substrate properties

Cells interact with their environment what manifests in the morphological, structural, and also mechanical alterations. The interaction is cell– and substrate-type specific. In one of the first papers [67], the effect of substrate properties was studied in the context of surface suitability in tissue engineering. Osteoblasts cultured on various substrates (CoCr, Ti, TiV, glass and tissue culture polystyrene) revealed the elasticity range from 2 Pa (observed for CoCr and TiV substrates) to 9 kPa for Ti surface. The latter modulus was comparable for that obtained for osteoblasts cultured in polysterene surface used for cell culture [67]. These results were related to the non-specific cell adhesion.

There are many studies focused on the role of extracellular matrix in cell functioning, including cancer progression [68, 69]. To assess the role of integrins in the interaction with ECM, glass or plastic surfaces are coated with the ECM protein, like fibronectin or collagen. Next, cells are cultured on such prepared surfaces and elastic properties are measured. Various results show that depending on the type of the ECM proteins cells exhibit distinct mechanical properties. One example of such studies has been reported by Docheva et al. [64]. The elastic properties and cytoskeleton organization of prostate cancer cells (less invasive LNCaP and more invasive PC-3 cell lines) have been studied in search of the answer whether collagen I produced in bones is responsible for cancer invasion and metastasis to bone. The PC-3 cells changed their elastic properties when they were cultured on surfaces coated with collagen I, what was accompanied by alterations in the cytoskeletal actin network [64]. The change of mechanical properties in the presence of ECM proteins is observed also for other cell types, e.g., for bladder cancer cells (Fig. 4.26).

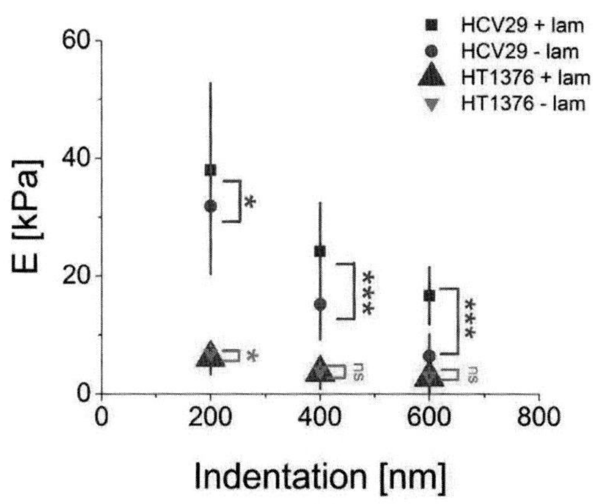

Figure 4.26 Elastic modulus to indentation depth relations, determined for human bladder cells (HCV29—non-malignant cell cancer of ureter, and HT1376—transitional cell carcinoma) cultured on a glass surface without and with laminin coating. Stars indicate statistical significance carried out using the Student *t*-test. Unpublished data of the author.

The results presented in Fig. 4.26 show clearly that the presence of laminin (one of the ECM proteins) changes mechanical properties of bladder cells, in particular, in the case of non-malignant HCV29 ones. The effect was independent on the indentation depth chosen for the analysis. Surprisingly, the presence of laminin is not influencing the cancerous HT1376 cells, probably indicating their large capability to adapt to various conditions.

Recently, the analysis of mechanical properties of cells as a function of surface properties started to be applied in the studies on the influence of extracellular matrix stiffness and structure on properties of cancerous cells aiming at the understanding the ECM role in cancer progression [65].

4.2.7.7 Comparing properties of human bladder cancer cells

At present, generally acknowledged opinion states that elasticity of cells cannot be considered in terms of absolute value of Young's modulus. Since mechanical properties of cells are dependent on so many various factors, such as cell density, number of passages and a day of measurements after the passage, medium composition, speed of force load, substrate chemical properties, probing tip geometry, the way how the recorded data are analyzed (apart the contact point determination), etc., comparison to the reference sample is essential. Table 4.2 summarizes the main experimental and analytical differences between mentioned above human bladder cell lines, HCV29 and T24 [38, 66, 67].

Table 4.2 Summary of main AFM experimental and analytical conditions set during elasticity measurements for human bladder cells (HCV29 and T24)

Cell type	v [μm/s]	Indentation [nm]	Surface	AFM probe	Model applied	E [kPa]
1999 [38]						
HCV29	0.19	600	PLL	MLCT (R = 50 nm)	P	14.8 ± 3.2
T24	0.19	1600	PLL	MLCT (R = 50 nm)	P	0.9 ± 0.4
2012 [66]						
HCV29	7	500	glass	MLCT (α = 35°C)	C	3.19 ± 0.27
T24	7	500	glass	MLCT (α = 35°C)	C	0.77 ± 0.25
2014 [67]						
HCV29	5	500	glass	MLCT (α = 20°C)	C	10 ± 3
T24	5	500	glass	MLCT (α = 20°C)	C	3.6 ± 2.1

Note: Culture medium composition (RPMI 1640 supplemented with 10% fetal bovine serum) was the same in all measurements.

Abbreviations: PLL, poly-L-lysine coated glass surface; glass, bare glass surface without PLL coating; P, the Hertz model with paraboloidal approximation of the AFM probe shape; C, the Hertz model with conical approximation.

In this particular case, large differences originate from all uncertainty sources, including cell-related factors, experimental conditions, data analysis, surface properties, etc. In the past, cell-related parameters (such as confluence on a substrate, the number of passages, the day of measurement after the passage, the cytoskeleton organization) were not well-controlled. Cells were considered ready for measurement when they formed a monolayer on poly-L-lysine coated glass coverslips. Later on, more focus was put into standardization of cells for the AFM measurements by adjusting the density of cells, number of passages and time of measurements after a passage, as well as by controlling surface properties. Moreover, the organization of actin cytoskeleton was visualized each time after the AFM experiments. Together with careful sample preparation, also experimental parameters provided by the AFM device should be also defined as well as possible. In 1999, the load speed was 190 nm/s, while in later experiments cells were measured at the speed of 5 and 7 μm/s. Finally, as it has been shown above, the geometry of the AFM also affects the cellular elasticity.

Despite above-mentioned drawbacks in the AFM-based absolute elasticity measurements, it should be underlined that the general relation between non-malignant and cancerous bladder cells remains the same as in the early studies, i.e., cancer cells are more deformable. This makes the AFM suitable for this kind of studies under the condition of using reference cells that must always be measured together with the studied material. Such approach allows proper normalization of the data (Fig. 4.27).

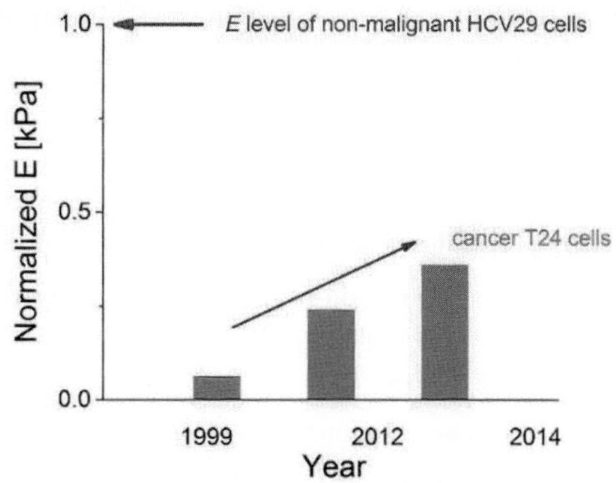

Figure 4.27 Normalized Young's modulus showing the mutual relation between non-malignant (HCV29) and cancerous (T24) cells, measured in different, independent experiments. The observed linear increase is totally accidental due to distinct experimental conditions applied in the studies. This image was prepared from data from Table 4.2.

Tuning various conditions of the elasticity measurements has a potential to emphasize or understate particular relative differences in cellular elastic properties. Obviously, more systematic studies on factors influencing mechanical properties of living cells will deliver more understanding on cell biomechanics and also will bring AFM toward a better detecting tool.

References

1. Ben-Ze'ev, A. (1997). Cytoskeletal and adhesion proteins as tumor suppressors. *Curr. Opin. Cell Biol.*, **9**, pp. 99–108.
2. Katira, P., Bonnecaze, R. T., and Zaman, M. H. (2013). Modelling the mechanics of cancer: Effect of changes in cellular and extracellular mechanical properties. *Front. Oncol.*, **3**(145), pp. 1–7.
3. Fung, Y. C. (1984). Structure and stress–strain relationship of soft tissues. *Amer. Zool.*, **24**, pp. 13–22.
4. Timoshenko, S., and Goodier, J. N. (1951). *Theory of Elasticity*. 2nd ed., McGraw-Hill Book Company, Inc., USA.
5. Yu, M. H., Ma, G. W., Qiang H. F., and Zhang, Y. Q. (2006) *Generalized Plasticity*, Springer-Verlag Berlin Heidelberg, New York.
6. Newman, J. (2008) *Physics of the Life Sciences,* Springer Science+ Business Media, LLC.
7. Quintana, J. M., Califano, A. N., Zaritzky, N. E., Partal, P., and Franco, J. M. (2002). Linear and nonlinear viscoelastic behavior of oil-in-water emulsions stabilized with polysaccharides. *J. Texture Studies,* **33** pp. 215–236.
8. Pipkin, A. C., and Rogers, T. G. (1968). A non-linear representation for viscoelastic behavior, *J. Mech. Phys. Solids,* **16**, pp. 59–72.
9. Kasza, K. E., Rowat, A. C., Liu, J., Angelini, T. E., Brangwynne, C. P., Koenderink, G. H., and Weitz, D. A. (2007). The cell as a material. *Curr. Opin. Cell Biol.*, **19**, pp. 101–107.
10. Thoumine, O., and Ott, A. (1997). Time scale dependent viscoelastic and contractile regimes in fibroblasts probed by microplate manipulation, *J. Cell Sci.*, **110**, pp. 2109–2116.
11. Bausch, A. R., Hellerer, U., Essler, M., Aepfelbacher, M., and Sackmann, E. (2001). Rapid stiffening of integrin receptor-actin linkages in endothelial cells stimulated with thrombin: A magnetic bead microrheology study, *Biophys. J.*, **80**, pp. 2649–2657.

12. Chen, S., Fatemi, M., and Greenleaf, J. F. (2004). Quantifying elasticity and viscosity from measurement of shear wave speed dispersion. *J. Acoust. Soc. Am.*, **115**, pp. 2781–2785.
13. Ferry, J. D. (1980). *Viscoelastic Properties of Polymers,* 4th ed., John Wiley & Sons Inc, USA.
14. Dong, C., Skalak, R., and Sung, K. L. (1991). Cytoplasmic rheology of passive neutrophils. *Biorheology*, **28**, pp. 557–556.
15. Ragsdale, G. K., Phelps, J., and Luby-Phelps, K. (1997). Viscoelastic response of fibroblasts to tension transmitted through adherens junctions. *Biophys. J.*, **73**, pp. 2798–2808.
16. Gautieri, A., Vesentini, S., Readaelli, A., and Buehler, M. J. (2012). Viscoelastic properties of model segments of collagen molecules. *Matrix Biol.*, **31**, pp. 141–149.
17. Monsia, M. (2011). A simplified nonlinear generalized maxwell model for predicting the time dependent behavior of viscoelastic materials. *World J. Mech.*, **1**, pp. 158–167.
18. Miller, K., and Chinzei, K. (1997). Constitutive modeling of brain tissue: Experiment and theory. *J. Biomech.*, **30**, pp. 1115–1121.
19. Chen, D. T. N, Wen, Q., Janmey, P. A., Crocker, J. C., and Yodh, A. G. (2010). Rheology of soft materials. *Annu. Rev. Condens. Matter. Phys.*, **1**, pp. 301–322.
20. Fielding, S. M., Sollich, P., and Cates, M. E. (2000). Aging and rheology in soft materials. *J. Rheol.*, **44**, pp. 323–369.
21. Herschel, W. H., and Bulkley, R. (1926), Konsistenzmessungen von Gummi-Benzollösungen. *Kolloid Zeitschrift,* **39**, pp. 291–300.
22. Bingham, E. C. (1916). An investigation of the laws of plastic flow. *U.S. Bur. Stand. Bull.*, **13**, pp. 309–353.
23. Fabry, B., Maksym, G., Butler, J., Glogauer, M., Navajas, D., and Fredberg, J. (2002). Scaling the microrheology of living cells. *Phys. Rev. Lett.*, **87**, paper no 148102, pp. 1–4.
24. Sollich, P. (1998). Rheological constitutive equation for a model of soft glassy materials. *Phys. Rev. E*, **58**, pp. 738–759.
25. Bursac, P., Lenormand, G., Fabry, B., Oliver, M., Weitz, D., Viasnoff, V., Butler, J., and Fredberg, J. (2005). Cytoskeletal remodeling and slow dynamics in the living cell. *Nat. Mater.*, **4**, pp. 557–561.
26. Deng, L., Trepat, X., Butler, J., Millet, E., Morgan, K., Weitz, D., and Fredberg, J. (2006). Fast and slow dynamics of the cytoskeleton. *Nat. Mater.*, **5**, pp. 636–640.
27. Fuller, B. (1961). Tensegrity. *Portfolio Artnews. Ann.*, **4**, pp. 112–127.

28. Ingber, D. E. (2008). Tensegrity-based mechanosensing from macro to micro. *Prog. Biophys. Mol. Biol.*, **97**, pp. 163–179.

29. Ingber, D. E. (2003). Tensegrity I. Cell structure and hierarchical systems biology. *J. Cell Sci.*, **116**, pp. 1157–1173.

30. Wang, N., Naruse, K., Stamenovic, D., Fredberg, J. J., Mijailovich, S. M., Tolic-Nørrelykke, I. M., Polte, T., Mannix, R., and Ingber, D. E. (2001). Mechanical behavior in living cells consistent with the tensegrity model. *Proc. Natl. Acad. Sci. U. S. A.*, **98**, pp. 7765–7770.

31. Wang, N., and Stamenovic, D. (2000). Contribution of intermediate filaments to cell stiffness, stiffening, and growth. *Am. J. Physiol Cell Physiol.*, **279**, pp. 188–194.

32. Stamenovic, D. (2008). Rheological behavior of mammalian cells. *Cell. Mol. Life Sci.*, **65**, pp. 3592–3605.

33. Canadas, P. V., Laurent, M., Oddou, C., Isabey, D., Wendling, S. (2002). A cellular tensegrity model to analyze the structural viscoelasticity of the cytoskeleton. *J. Theor. Biol.*, **218**, pp. 155–173.

34. Butcher, D. T., Alliston, T., and Weave, V. M. (2009). A tense situation: Forcing tumor progression. *Nat. Rev. Can.*, **9**, pp. 108–122.

35. Briscoe, B. J., Fiori, L., and Pelillo, E. (1998). Nano-indentation of polymeric surfaces. *J. Phys. D Appl. Phys.*, **31**, pp. 2396–2405.

36. Pasternak, C., Wong, S., and Elson, E. L. (1995). Mechanical function of dystrophin in muscle cells. *J. Cell Biol.*, **128**, pp. 355–361.

37. Paszek, M. J., Zahir, N., Johnson, K. R., Lakins, J. N., Rozenberg, G. I., Gefen, A., Reinhart-King, C. A., Margulies, S. S., Dembo, M., Boettiger, D., Hammer, D. A., and Weaver, V. M. (2005). Tensional homeostasis and the malignant phenotype. *Cancer Cell*, **8**, pp. 241–254.

38. Lekka, M., Laidler, P., Gil, D., Lekki, J., Stachura, Z., and Hrynkiewicz, A. Z. (1999). Elasticity of normal and cancerous human bladder cells studied by scanning force microscopy. *Eur. Biophys. J.*, **28**, pp. 312–316.

39. Lim, C. T., Zhou, E. H., and Quek, S. T. (2004). Mechanical models for living cells—a review. *J. Biomech.*, **39**, pp. 195–216.

40. Guck, J., Lautenschläger, F., Paschke, S., and Beil, M. (2010). Critical review: Cellular mechanobiology and amoeboid migration. *Integr. Biol.*, **2**, pp. 575–583.

41. Hochmuth, R. M. (2000). Micropipette aspiration of living cells. *J. Biomech.*, **33**, pp. 15–22.

42. Thoumine, O., and Ott, A. (1997). Time scale dependent viscoelastic and contractile regimes in fibroblasts probed by microplate manipulation. *J. Cell Sci.*, **110**, pp. 2109–2116.

43. Mao, X., and Huang, T. J. (2012). Exploiting mechanical biomarkers in microfluidic. *Lab on Chip*, **12**, pp. 4006–4009.

44. Guck, J., Schinkinger, S., Lincoln, B., Wottawah, F., Ebert, S., Romeyke, M., Lenz, D., Erickson, H. M., Ananthakrishnan, R., Mitchell, D., Käs, J., Ulvick, S., and Bilby, C. (2005). Optical deformability as an inherent cell marker for testing malignant transformation and metastatic competence. *Biophys. J.*, **88**, pp. 3689–3689.

45. Qian, H., Sheetz, M. P., Elson, E. L. (1991). Single particle tracking. Analysis of diffusion and flow in two-dimensional systems. *Biophys. J.*, **60**, pp. 910–921.

46. Puig-De-Morale, M., Grabulosa, M., Alcaraz, J., Mullol, J., Maksym, G. N., Fredberg. J. J., and Navajas, D. (2001). Measurement of cell microrheology by magnetic twisting cytometry with frequency domain demodulation. *J. Appl. Physiol.*, **91**, pp. 1152–1159.

47. Hertz, H. (1881). Über die Berührung fester elastischer Körper, Journal für die reine und angewandte. *Mathematik*, **92**, pp. 156–171.

48. Sneddon, I. N. (1965). The relation between load and penetration in axisymetric Boussinesq problem for a punch of arbitrary profile. *Int. J. Engng. Sci.*, **3**, pp. 47–57.

49. www.jpk.org.

50. Weisenhorn, A., Khorsandi, M., Kasas, S., Gotzos, V., and Butt, H. J. (1993). Deformation and height anomaly of soft surfaces studied with an AFM. *Nanotechnology*, **4**, pp. 106–113.

51. Pogoda, K., Jaczewska, J. Wiltowska-Zuber, J. Klymenko, O., Zuber, K., Fornal, M., and Lekka, M. (2012). Depth-sensing analysis of cytoskeleton organization based on AFM data. *Eur. Biophy. J.*, **41**, pp. 79–87.

52. Roduit, C., Sekatski, S., Dietler, G., Catsicas, C., Lafont, F., and Kasas, S. (2009). Stiffness tomography by atomic force microscopy. *Biophys. J.*, **97**, pp. 674–677.

53. Longo, G., Marques Rio, L., Roduit, C., Trampuz, A., Bizzini, A., Dietler, G., and Kasas, S. (2012). Force volume and stiffness tomography investigation on the dynamics of stiff material under bacterial membranes. *J. Mol. Recognit.*, **25**, pp. 278–284.

54. Li, Q. S., Lee, G. Y., Ong, C. N., and Lim, C. T. (2008). AFM indentation study of breast cancer cells. *Biochem. Biophys. Res. Commun.*, **374**, pp. 609–613.

55. Lekka, M., Gil, D., Pogoda, K., Dulińska-Litewka, J., Jach, R., Gostek, J., Klymenko, O., Prauzner-Bechcicki, Sz., Stachura, Z., Wiltowska-

Zuber, J., Okoń, K., and Laidler, P. (2012) Cancer cell detection in tissue sections using AFM. *Arch. Biochem. Biophys.*, **518,** pp. 151–156.

56. Faria, E. C., Ma, N., Gazi, E., Gardner, P., Brown, M., Clarke, N. W., and Snook, R. D. (2008). Measurement of elastic properties of prostate cancer cells using AFM. *Analyst,* **133**, pp. 1498–1500.
57. Ketene, A. N., Schmelz, E. V., Roberts, P. C., and Agah, M. (2012). The effects of cancer progression on the viscoelasticity of ovarian cell cytoskeleton structures. *Nanomed. Nanotechnol. Biol. Med.,* **8**, pp. 93–102.
58. Prabhune, M., Belge, G., Dotzauer, A., Bullerdiek, J., and Radmacher, M. (2012). Comparison of mechanical properties of normal and malignant thyroid cells. *Micron*, **43**, pp. 1267–1272.
59. Nikkhah, M., Strobl, J. S., Schmelz, E. M., and Agah, M. (2011). Evaluation of the influence of growth medium composition on cell elasticity. *J. Biomech.,* **44**, pp. 762–766.
60. Guo, X., Bonin, K., Scarpino, K., and Guthold M. (2014). The effect of neighbouring cells on the stiffness of cancerous and non-cancerous human mammary epithelial cells. *New J. Phys.,* **16**, pp. 105002.
61. Chio, Y. W., Lin, H. K., Tang, M. J., Lin, H. H., and Yeh, M. L. (2013). The influence of physical and physiological cues on atomic force microscopy-based cell stiffness assessment. *PLoS ONE,* **8**, pp. e77384.
62. Zhao, M., and Huey, B. D. (2006). Rate- and depth-dependent nanomechanical behavior of individual living Chinese hamster ovary cells probed by atomic force microscopy. *J. Mater. Res.,* **21**, pp. 1906–1911.
63. Domke, J., Dannohl, S., Parak, W. J., Mueller, O., Aicher, W. K., and Radmacher, M. (2000). Substrate dependent differences in morphology and elasticity of living osteoblasts investigated by atomic force microscopy. *Colloids Surfaces B,* **19**, pp. 367–379.
64. Docheva, D., Padula, D., Schieker, M., and Clausen-Schaumann, H. (2010). Effect of collagen I and fibronectin on the adhesion, elasticity and cytoskeletal organization of prostate cancer cells. *Biochem. Biophys. Res. Commun.,* **12**, pp. 361–366.
65. HBaker, E. L., Bonnecaze, R. T., and Zaman, M. H. (2009). Extracellular matrix stiffness and architecture govern intracellular rheology in cancer. *Biophys. J.,* **97**, pp. 1013–1021.

66. Lekka, M., Pogoda, K., Gostek, J., Klymenko, O., Prauzner-Bechcicki, Sz., Wiltowska-Zuber, J., Jaczewska, J., Lekki, J., and Stachura, Z. (2012). Cancer cel recognition—mechanical phenotype. *Micron,* **43**, pp. 1259–1266.

67. Ramos, J. R., Pabijan, J., Garcia, R., and Lekka, M. (2014). The softening of human bladder cancer cells happens at an early stage of the malignancy processes. *Beilstein J. Nanotechnol.,* **5**, pp. 447–457.

Chapter 5

Adhesive Properties Studied by AFM

Biological adhesion, at cellular level, is realized by the formation of adhesive contacts, linking ligands to surface receptors. A contact between a single ligand and receptor molecules is relatively weak and not sufficient to maintain a whole cell adhesion to neighboring cells or to environment provided by ECM. In nature, a large number of adhesion receptors, together with their dense arrangement on a cell membrane, facilitate the formation of stable and long-living adhesion sites using individual weak bonds. The existence of many weak bonds instead of a single strong bond probably enables the cell to gain the flexibility and capability for rapid rearrangement in response to external stimuli [1]. Moreover, single molecular complexes, usually, have to operate under the presence of external forces. Thus, to understand the cell adhesion, it is crucial to investigate a bond failure under certain conditions. Such studies have become valuable for biologists since they can deliver information about the structure of adhesion molecules and their binding/unbinding properties. In particular, if such measurements are carried out on a surface of living cells in conditions close to natural ones, the better understanding of molecules' functioning in their normal conditions might be obtained [2]. This chapter briefly introduces the basis for the force-induced unbinding experiments carried out by atomic force microscopy.

Cellular Analysis by Atomic Force Microscopy
Malgorzata Lekka

ISBN 978-981-4669-67-2 (Hardcover), 978-1-315-36480-3 (eBook)
www.panstanford.com

5.1 Unbinding of Molecules: Theoretical Basis

5.1.1 Brief Introduction to Kramer's Theory

The bond dissociation[1] (or molecular complex unbinding) is usually approximated by a particle moving over an energy barrier of a one-dimensional energy landscape describing the binding pocket of two interacting molecules [3]. Figure 5.1 schematically presents one-dimensional energy landscape as a model for a single molecular bond or complex interaction as a function of a reaction coordinate x. The bound state (characterized by a deep energy minimum ΔG_0 appearing at small distances, usually set as "zero") is separated from the unbound one by a transition state (ΔG_a) represented by an energy barrier located at the distance x_b.

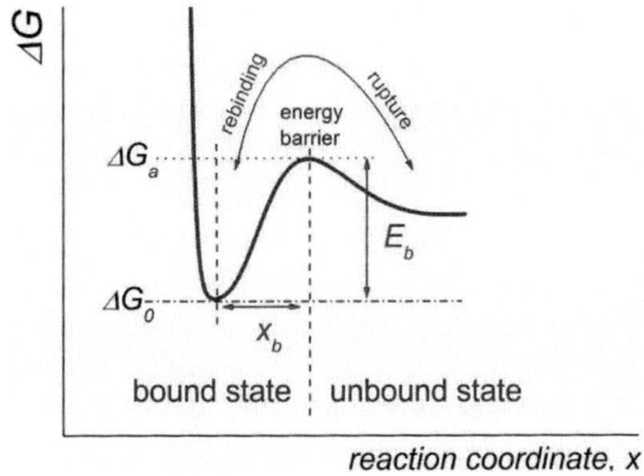

Figure 5.1 Scheme of one-dimensional energy landscape of the unbinding process of two interacting molecules as a function of the reaction coordinate x.

Initially, the unbinding process has been approximated as a chemical reaction of two reactants M_1 and M_2, described by two reaction rates k_{on} and k_0, corresponding to the rate of complex formation and its dissociation, respectively:

$$M_1 \underset{k_0}{\overset{k_{on}}{\Leftrightarrow}} M_2 \tag{5.1}$$

---

[1]Here, the term "single bond" refers to "single molecular complex", under assumption that rupture of a single molecular complex can be approximated by the rupture of a single bond.

This relation is assumed to be a first-order kinetics because the rate of change depends linearly on the concentration of reactants, like in various processes involving proteins [4]. It is described by the following formula:

$$\frac{dM_1}{dt} = -k_{on}M_1 + k_0 M_2 \tag{5.2}$$

Assuming that transition proceeds through an energy barrier, the rate for a barrier crossing (k_0) is exponentially dependent on both the height of the energy barrier and the temperature. Later on, it was realized that escape from a bound state could happen only via noise-assisted hopping events, governed by the theory of Brownian motion (i.e., the rupture and rebinding events correspond to transitions of the Brownian particle over the energy barrier [5]). In the Kramer's theory, the rate of escape from the state of a local stability (i.e., a bound state) along a preferential path over the barrier is driven by thermal forces F, dependent on temperature T and the frequency of attempts γ_D. The expression for the dissociation rate is

$$k_o = \gamma_D \cdot e^{-\frac{\Delta G_a - \Delta G_0}{k_B T}} = \gamma_D \cdot e^{-\frac{E_b}{k_B T}}, \tag{5.3}$$

where $E_b = \Delta G_a - \Delta G_0$ is the height of the barrier relative to the bound state, and k_B is Boltzmann's constant. The dissociation rate constant k_0 is the inverse of the bond lifetime, expressed by

$$\tau_0 = \frac{1}{\gamma_D} \cdot e^{\frac{E_b}{k_B T}} \tag{5.4}$$

In the transition state theory, a prefactor γ_D is the natural vibration frequency of the bond and is of the order of $\gamma_D \sim 5 \times 10^{13}\ s^{-1}$ for C–C bonds [6, 7]. If the thermal energy $k_B T$ is much smaller than the respective barrier height, the forces act as a small perturbation, and the system stays at the deep minimum for a long time (there is only a small probability that it accumulates enough energy to overcome the energy barrier to reach the unbound state). If the thermal energy is comparable with or even larger than the barrier height, the particle can move almost freely from the bound to unbound states. Thermal fluctuations can be large enough to overcome more than one energy barrier between bound and unbound states [7, 8].

5.1.2 Force-Induced Single Bond Disruption

In many biochemical processes, the bond rupture is induced by external forces. According to the transition state theory [5], the external force tilts the energy landscape, i.e., decreases the energy barrier height (Fig. 5.2).

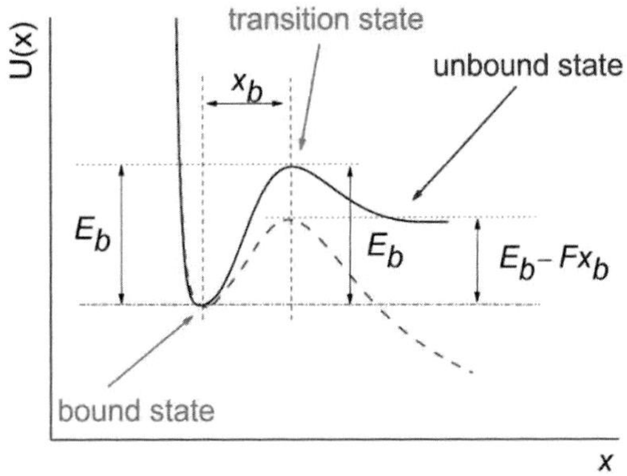

Figure 5.2 Schematic representation of the energy landscape of two interacting molecules (i.e., ligand–receptor complex, black line). The applied external force F (acting e.g., during measurements by means of atomic force microscopy) lowers the height of the energy barrier, resulting in easier bond disruption (dashed line). The transition state is characterized by a barrier height E_b and its distance from the bound state x_b. Reprinted with permission from [2].

The Kramer's theory assumes that the motion of the ligand proceeds in the strong friction limit ($\eta >> 1$) along a reaction coordinate x and is governed by the one-dimensional Langevin's equation:

$$\eta \frac{\partial}{\partial t} x(t) = -\frac{\partial}{\partial x} U(x) + F(x,t) + \sigma \cdot N(t), \tag{5.5}$$

where $F(x, t)$ is the external force applied to the ligand and $\sigma N(t)$ is a stochastic force of an amplitude σ and a zero mean. One way to apply external forces to a protein–ligand complex is to remove a ligand (by pulling) from its initial position within the binding pocket. In case of a constant pulling velocity v and under the external potential described by $U(x) = k_{\text{complex}} \cdot (x - x_b)^2/2$, the external pulling force F can be expressed as a function of single reaction coordinate x:

$$F = k_{\text{complex}} \cdot (x_b + v \cdot t - x), \tag{5.6}$$

where k_{complex} is the stiffness of the ligand–receptor complex and x_b is the distance between the bound and unbound states. If the transition from bound to unbound states over the energy barrier is associated with displacement in the direction of the acting force, the height of the energy barrier is lowered by a factor Fx_b. This leads to the basic equation of the Bell model:

$$k_{\text{off}} = \gamma_D \cdot e^{-\frac{E_b - Fx_b}{k_B T}} = k_0 \cdot e^{\frac{Fx_b}{k_B T}} \tag{5.7}$$

The Bell model shows that rupture forces are dependent on the intrinsic lifetime of the bond, the temperature, and on the measurement time [7]. It is valid if the energy landscape under applied force is the sum of the original potential (in the absence of force) and the (linear) potential defined by the applied force with a fixed distance x_b between the bound and unbound states along the reaction coordinate and unbinding path.

Equation (5.7) does not account for the stochastic nature of a single bond rupture, arising from random fluctuations of the system in its equilibrium state. In Bell's theory, the unbinding proceeds through the single reaction pathway defined by the presence of the external force. To introduce stochastic nature of the unbinding, Evans and Ritchie proposed a model to describe force-induced unbinding that accounts for the experimentally observed distributions of measured unbinding forces [9]. From the first-order of the kinetic theory [6, 10], for the irreversible unbinding, the probability that a single complex breaks in the time interval $(t, t + dt)$ is

$$p(t + dt) = p(t) \cdot (1 - k_{\text{off}}(t) \cdot dt) \tag{5.8}$$

and therefore,

$$\frac{dp(t)}{dt} = -k_{\text{off}}(t)p(t) \qquad (5.9)$$

Taking into account Eq. (5.7), the probability density function $p(t, F)$ of a single dissociating complex in the time interval $(t, t + dt)$ as a function of force, F, is

$$p(t,F) = k_{\text{off}}(F(t)) \cdot e^{-\int_0^t k_{\text{off}}(F(t'))dt'}. \qquad (5.10)$$

The first term represents the probability of dissociation in the next short time interval, dt, whereas the exponential term describes the probability of the complex survival in the bound state up to time t. The solution of Eq. (5.10) is frequently derived under the assumptions that the pulling force changes linearly in time:

$$F(t) = k_{\text{syst}} \cdot v \cdot t, \qquad (5.11)$$

where k_{syst} is the effective spring constant accounting for the AFM cantilever and the single molecular complex spring constants, and v is the tip retraction velocity. In that case, the probability distribution is expressed by the following equation (normalized to 1):

$$p(t,F) = \frac{k_0}{r_{\text{f}}} \cdot e^{\frac{x_{\text{b}} \cdot F}{k_{\text{B}} \cdot T}} \cdot e^{-\frac{k_0 \cdot k_{\text{B}} \cdot T}{x_{\text{b}} \cdot r_{\text{f}}}\left(e^{\frac{x_{\text{b}} \cdot F}{k_{\text{B}} \cdot T}} - 1\right)}, \qquad (5.12)$$

where $r_{\text{f}} = k_{\text{syst}} \cdot v$ is the loading rate, describing how fast the force changes in time during molecular complex rupture. The first experiments reporting the unbinding of single molecular complexes were carried out mostly at fixed pulling velocities and spring constants. More experiments shown that the rupture force of an isolated complex is dependent on the force loading rate. This has been shown in several experiments by employing the dynamic force spectroscopy (DFS) technique, where the applied force-loading rate was varied over a few orders of magnitude [11–13]. Equation (5.12) shows that the probability distribution of the unbinding force depends on the loading rate r_{f} (Fig. 5.3).

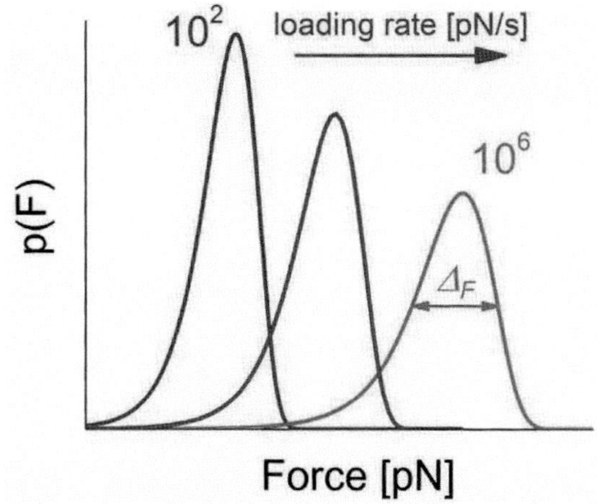

Figure 5.3 Probability distributions as a function of an unbinding force calculated for different loading rates (from 100 to 10^6 pN/s).

The larger loading rate shifts the maximum of the distribution towards larger unbinding force values and, in addition, broadens the force distribution. The width of the distribution is given by the following equation:

$$\frac{1}{\Delta_F^2} = \left[-\left(\frac{1}{p(F)} \right) \cdot \frac{\partial^2 p(F)}{\partial F^2} \right] \tag{5.13}$$

The maximum of the distribution (the most probable unbinding force), defined by the condition $dp(F)/dF = 0$ is located at

$$F_{\text{unb}}(r_{\text{f}}) = \frac{k_{\text{B}} \cdot T}{x_{\text{b}}} \cdot \ln\left(\frac{x_{\text{b}} \cdot r_{\text{f}}}{k_0 \cdot k_{\text{B}} \cdot T} \right). \tag{5.14}$$

Equation (5.14) can be rewritten as follows [9, 14]:

$$F_{\text{unb}} = \frac{k_{\text{B}} \cdot T}{x_{\text{b}}} \cdot \ln\left(\frac{x_{\text{b}}}{k_0 \cdot k_{\text{B}} \cdot T} \right) + \frac{k_{\text{B}} \cdot T}{x_{\text{b}}} \ln(r_{\text{f}}) \tag{5.15}$$

This form clearly shows the linear dependence of the most probable unbinding force of a single molecular complex on the logarithm of loading rate $\ln(r_{\text{f}})$.

The example of such study is shown in Fig. 5.4, for the molecular complex composed of albumin and its monoclonal antibody (BSA–aBSA complex).

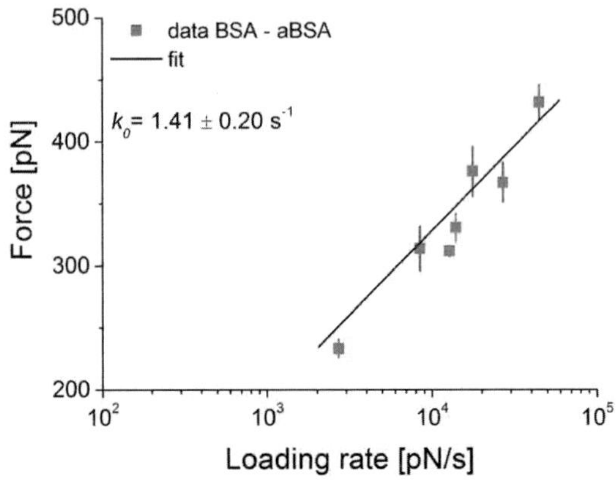

Figure 5.4 The dependence of the most probable unbinding force on the logarithm of loading rate measured for the interaction between albumin (BSA) and its antibody (aBSA). Reprinted with permission from [2].

The obtained relation was fitted with a single line, showing that unbinding of these molecules possessed through a single energy barrier. The corresponding dissociation rate constant was of 1.41 ± 0.20 s^{-1}, corresponding to the bond lifetime of 0.71 s. Such a linear dependence of the most probable unbinding force on the logarithm of the loading rate has been commonly observed for most of antigen–antibody interactions [2, 10–13].

5.1.3 Hierarchic Crossing through the Energy Barriers

For almost all molecular complexes investigated so far, the dependence between the unbinding force and the logarithm of the loading rate is represented by a single line, indicating the presence of only one energy barrier in the interaction energy landscape. However, the unbinding that involves single molecules with complex unbinding sites may exhibit multiple local maxima and minima in the interaction potential along the reaction coordinate. In such cases, the plot of the most probable unbinding force and logarithm of the loading rate shows a sequence of lines with different slopes, each corresponding to the position of a particular energy barrier. Apart from biotin–(strept)avidin complex [15, 16], the hierarchic crossing through energy barriers has been shown, for instance, for individual $\alpha 5\beta 1$ integrin–fibronectin complexes [17]. Another example showing two linear regions in the dependence of the unbinding force on the loading rate is presented in Fig. 5.5.

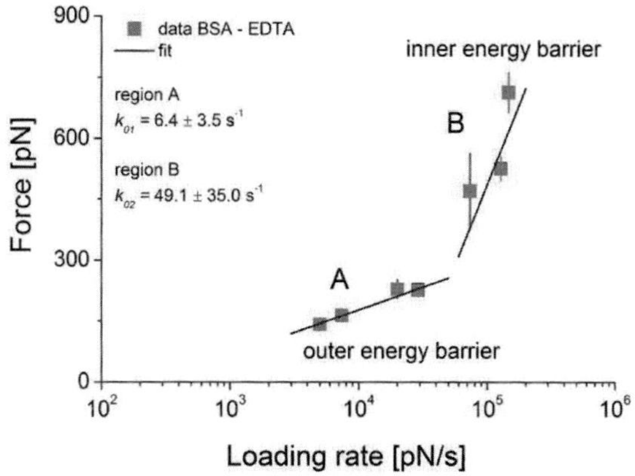

Figure 5.5 (A) The force versus loading rate dependence measured for BSA–EDTA complex showing two segments of linear dependencies corresponding to two energy barriers (outer and inner one). Reprinted with permission from [2].

The measurements were carried out for two molecules: albumin and ethylene-di-amine tetraacetic acid (EDTA). The observed two linear segments indicated existence of two barriers (inner and outer ones) in the corresponding energy landscape. For each of them a separate set of the Bell–Evans model parameters could be determined. The region A (Fig. 5.5) corresponds to the outermost energy barrier characterized by the dissociation rate constant of k_{01} = 6.4 ± 3.5 s^{-1}. Passing through inner energy barrier (region B, Fig. 5.5) is described by the dissociation rate constant of k_{02} = 49.1 ± 35.0 s^{-1}. The extrapolation of the different linear segments to F = 0 differs by the amount related to the relative differences in the magnitude of the individual energy (the absolute values of multiple activation barriers can no longer be calculated directly). During the unbinding, the complex stays for 0.020 s in the first transition state (inner barrier) and for 0.156 s in the second transition state (outer barrier).

From the theory, for a hierarchic crossing, each *i*-th barrier is described by its own dissociation rate constant (k_0^i) and a position of the energy barrier (x_b^i). In such case, the force-induced unbinding is modeled as the crossing over a series of barriers [18]. Then, the overall unbinding rate (k_{off}) is a sum of a specific dissociation rates:

$$\frac{1}{k_{\text{off}}(F)} = \sum_{i=1}^{N} \frac{1}{k_{\text{off}}^i(F)} \tag{5.16}$$

When the system must overcome a series of activation barriers, the dissociation kinetics at low loading rate is dominated by properties of the outermost barrier. With increase of the loading rate, the outermost barriers are suppressed, and the dissociation is governed by the inner activation barriers [18].

5.1.4 The Energy Barrier Height

In the Bell–Evans model applied to the AFM measurements only the position of the energy barrier and the dissociation rate constant can be estimated. It is difficult to quantify the absolute value of the energy barrier height without involving other techniques, such as quartz microbalance [19], surface plasmon resonance [20] or calorimetry [21]. However, despite the lack of the absolute value of the barrier heights, the relative difference between barriers can be estimated by knowing dissociation rate constants according to

$$\Delta(\Delta G) = \Delta G_1 - \Delta G_2 = k_B T \cdot \ln\left(\frac{k_{02}}{k_{01}}\right), \tag{5.17}$$

where k_{01}, k_{02} are the dissociation rate constants determined for two different molecular complexes, under assumption that the association constant (k_{on}) remains unchanged during the complex formation. On the other hand, assuming that all molecules will react and knowing k_0, one can estimate ΔG as

$$-\Delta G = k_B T \cdot \ln\left(\frac{k_0 \cdot h}{k_B \cdot T}\right), \tag{5.18}$$

where h is Planck's constant [9].

5.1.5 Multiple Bond Rupture

The adhesion realized by single cells is expected to involve various types of ligand–receptor complexes, mediating contact with other cells or to the extracellular matrix, acting as a cluster of molecular complexes. The unbinding of such a cluster can proceed through more multifaceted pathways. The simplest theory assumes uncooperative nature of unbinding[2]. Then, two extreme cases can be considered i.e., a "*zipper-like*" or "*parallel-like*" models, denoting consecutive or simultaneous rupture of molecular complexes, respectively. In the "*zipper-like*" model, the whole external force is exerted on the first bond and as soon as it breaks, the next bond is loaded. This model is frequently applied in modeling of RNA or DNA unzipping [22, 23]. It can also be used for a sequential rupture of multiple bonds, as it takes place in leukocyte rolling [10, 11]. In the "*parallel-like*" model, the unbinding of several complexes occurs at the same time. As a consequence, the applied external force is the product of the number of ruptured single molecular complexes times the value of the force needed to break a single complex alone [24, 25].

5.1.5.1 Sequential bond rupture: the "zipper-like" model

In the "*zipper-like*" model (Fig. 5.6), the applied external force breaks each individual ligand–receptor complex separately, one after another, like in a zipper [24].

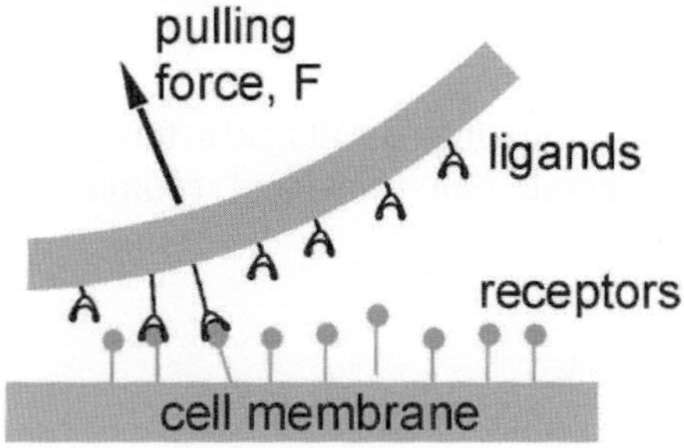

Figure 5.6 Schematic representation of the consecutive unbinding of individual complexes—the "*zipper-like*" model. Reprinted with permission from [2].

---

[2]Under assumption that force drops to zero, each individual molecular complex behaves as a time separated unbinding event, characterized by a single energy barrier potential with no dependence on the previous unbinding process.

In *"zipper-like"* model, each individual molecular complex unbinds under the same value of the external force, *F*. Also, the unbinding proceeds through the same single–barrier potential, described by the dissociation rate constant k_0 and by the position of energy barrier x_b:

$$k_{\text{off}}(F) = k_0 \cdot e^{\frac{x_b \cdot F}{k_B \cdot T}} \tag{5.19}$$

Thus, the overall dissociation rate of all identical *N* complexes is the following:

$$k_{\text{off}}^N(F) = \left(\sum_{i=1}^{N} \frac{1}{k_0} \cdot e^{-\frac{F \cdot x_b}{k_B \cdot T}} \right)^{-1} = \frac{k_0}{N} \cdot e^{\frac{x_b \cdot F}{k_B \cdot T}} \tag{5.20}$$

The most probable unbinding force for *N* bonds ruptured sequentially, F_{unb}^Z can be determined from the unbinding probability density (Eq. 5.10) by setting the condition $dp(F)/dF = 0$:

$$F_{\text{unb}}^Z(N) = \frac{k_B \cdot T}{x_b} \cdot \left(\ln(r_f) + \ln\left(\frac{x_b}{k_0 \cdot k_B \cdot T} \right) + \ln(N) \right). \tag{5.21}$$

Finally, for the single complex unbinding, one can obtain

$$F_{\text{unb}}^Z(N) = F_{\text{unb}} + \frac{k_B \cdot T}{x_b} \cdot \ln(N). \tag{5.22}$$

Here, zipper-like unbinding, results in a force value larger than that needed to break one individual complex. The difference is defined as

$$G = \frac{k_B \cdot T}{x_b} \cdot \ln(N). \tag{5.23}$$

The relation between its value and the number of bonds simultaneously ruptured is presented in Fig. 5.7. It shows that the unbinding force *F* increases with the increase of the energy barrier width x_b and the number of simultaneously ruptured single bonds (or single complexes), *N*.

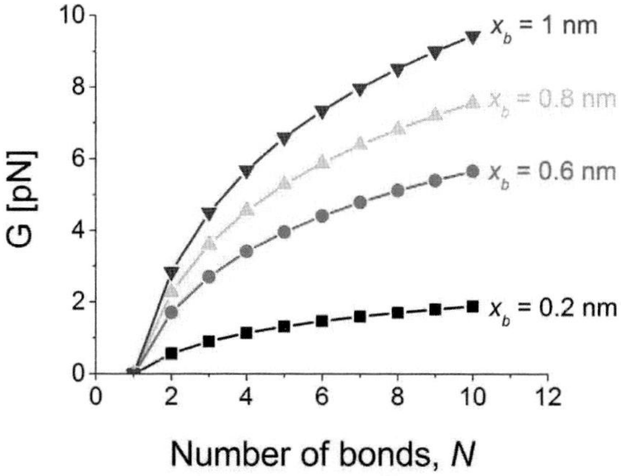

Figure 5.7 The correction factor (G) that has to be added to the unbinding force value in *"zipper-like"* model, plotted as a function of the number of simultaneously broken bonds/single complexes calculated for energy barrier positions ranged from 0.2 to 1 nm ($k_BT = 4.1$ pN·nm).

The measured x_b values are usually lower than 1 nm: For different types of molecular complexes, the position of the energy barrier has been reported to vary from 0.08 to 0.55 nm [13, 18, 26]. For x_b = 0.2 nm, the contribution of the G factor to the unbinding force can be neglected, since its value is lower than the thermal fluctuations of the cantilever (for example, for the spring constant of 0.01 N/m, the level of thermal fluctuations is 6.6 pN).

When the unbinding follows the *"zipper-like"* model, the distance between two subsequent maxima in the force histogram is

$$\frac{k_B \cdot T}{x_b} \cdot \ln\left[\frac{i+1}{i}\right]. \tag{5.24}$$

5.1.5.2 Sequential bond rupture: the "parallel-like" model

The theoretical basis of *"parallel-like"* model describing the simultaneous rupture of N–individual complexes has been presented by Williams [24]. During such unbinding, the external force is equally shared between all participating complexes. Then, the dissociation rate of rupture for all N identical uncooperative complexes is the following:

$$k_{\text{off}}^{N}(F)=\left(\sum_{n=1}^{N}\frac{k_0}{n}\cdot e^{\frac{-x_b\cdot F}{n\cdot k_B\cdot T}}\right)^{-1} \tag{5.25}$$

As a consequence, for large number of individual complexes, the unbinding force is approximated by

$$F_{\text{unb}}^{P}\approx N\cdot F+N\cdot\frac{k_B T}{x_b}\cdot\left(\ln\left(\frac{k_B T}{x_b}\right)-\ln(F_{\text{unb}}^{P})\right), \tag{5.26}$$

where $F_{\text{unb}}^{P} < N\cdot F$.

In the AFM measurements, when usually few molecular complexes are simultaneously ruptured (*N* is small), the measured unbinding force is a sum of those characteristic for the single complexes. Moreover, very often, the cooperative character of the unbinding is observed (Fig. 5.8).

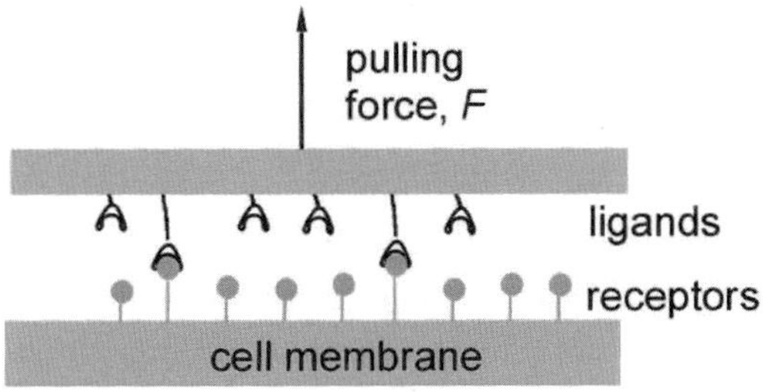

Figure 5.8 Scheme of the simultaneous rupture of cooperative *N*-bonds—the "*parallel-like*" model. Reprinted with permission from [2].

When multiple molecular complexes act as cooperative bonds, the unbinding rate for breaking of *N* bonds is the following:

$$k_{\text{off}}^{N}(F)=\frac{k_0}{N}\cdot e^{\frac{x_b\cdot F}{N\cdot k_B\cdot T}} \tag{5.27}$$

The probability density function $p(t, F)$ describing N single complexes dissociating in the time interval $(t, t + dt)$ is

$$p(F)=\frac{k_0}{N}\cdot e^{\frac{x_b\cdot F}{N\cdot k_B\cdot T}}\cdot e^{-\frac{k_0\cdot k_B\cdot T}{x_b\cdot r_f}\cdot\left(e^{\frac{x_b\cdot F}{N\cdot k_B\cdot T}}-1\right)} \tag{5.28}$$

The corresponding most probable unbinding force, F^P_{unb}:

$$F^P_{\text{unb}}(N) = \frac{N \cdot k_B \cdot T}{x_b} \cdot \left(\ln(r_f) + \ln\left(\frac{x_b}{k_0 \cdot k_B \cdot T} \right) \right) \tag{5.29}$$

Analogously, as for "*zipper-like*" model:

$$F^P_{\text{unb}}(N) = N \cdot F_{\text{unb}}. \tag{5.30}$$

Thus, many individual complexes unbind under the shared force, resulting in the force value being the product of the number of ruptured complexes and the force value needed to break one individual complex. The distance between two subsequent maxima, corresponding to the unbinding of the *i*-th and (*i* + 1)-th complex is equal to F_{unb}.

5.1.6 Comparing Unbinding Properties of Two Single Complexes

In majority of cases, the force-induced unbinding, studied for various molecular complexes, shows linear dependency between the unbinding force and logarithm of the loading rate [9]. Each of obtained lines can be characterized by a slope and an intercept, attributed to the width of the energy barrier (x_b) and the dissociation rate constant (k_0) by the following relations (from Eq. 5.12):

$$\text{Slope} = \frac{k_B \cdot T}{x_b} \tag{5.31}$$

$$\text{Intercept} = \frac{k_B \cdot T}{x_b} \cdot \ln\left(\frac{x_b}{k_0 \cdot k_B \cdot T} \right), \tag{5.32}$$

where k_B is Boltzmann's constant and T is the temperature (their product is 4.1 nm · pN at room temperature).

Having in mind that the unbinding force value measured through pulling experiments depends on the loading rate, the basic question is whether it is possible to compare the unbinding process for two distinct molecular complexes. Such comparison is possible only for cases when the unbinding proceeds at the same loading rate value. Moreover, knowing the relations between the unbinding force and the logarithm of the loading rate, one can guess the mutual relation between the energy landscape shape of these complexes. Figure 5.9 schematically illustrates three specific loading rate dependencies that can be obtained from the AFM unbinding experiments.

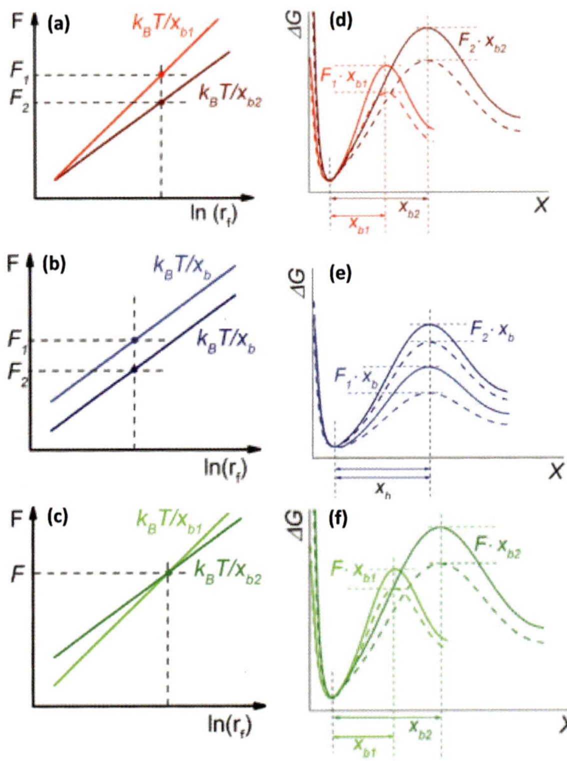

Figure 5.9 The schematic relations between the unbinding force and logarithm of the loading rate for two molecular complexes undergoing force-induced unbinding (a–c) and the corresponding energy landscapes (d–f).

In the most common scenario (Fig. 5.9a), two intersecting lines are usually observed. They indicate that the energy landscape shows only the outermost energy barrier. At the same loading rate value (vertical dashed line in Fig. 5.9a), two different values of the unbinding force F_1 and F_2, are observed for the *complex_1* and *complex_2*, respectively. Using the Bell–Evans model to determine parameters describing the single-molecule unbinding, the energy landscape can be reconstructed. Two different slopes, observed for each molecular complex, denote two different widths of the energy barrier (x_{b1} and x_{b2}). The intercept value for *complex_1* is smaller than for the *complex_2*, which indicates that the energy barrier height of the *complex_2* is larger than the energy barrier height of the *complex_1* of about $\Delta G = k_B T \ln(k_2/k_1)$ (according to Eq. 5.17).

In summary, the resulting energy landscape for the *complex_2* shows higher and wider energy barrier as compared to energy landscape of the *complex_1*. During force-induced unbinding, the applied external forces lower the energy barriers by about $F_1\, x_{b1}$ and $F_2\, x_{b2}$, correspondingly for *complex_1* and *complex_2*. Knowing the exact numbers of F and x_b, it is possible to determine the magnitude of the energy barrier drop for the studied complexes. This immediately enable to draw schematically how the energy landscape looks like for molecular complexes pulled apart at the same loading rate. Thus, two specific cases can be considered.

The first case assumes that the relations between the unbinding force and the loading rate are parallel, for two complexes under considerations (Fig. 5.9b). In this case, the slopes denote the same energy barrier width (x_b). Analogously, as above, the energy barrier height of the *complex_2* is larger than the energy barrier height of the *complex_1* since the intercept value for *complex_1* is smaller than for the *complex_2*. Moreover, knowing that unbinding force F_1 is larger than the unbinding force for the *complex_2*, it is easily seen that the applied external force F_1 reduces more the energy barrier height for *complex_1* than for *complex_2* ($F\, x_{b1} > F\, x_{b2}$).

The other specific case assumes that the unbinding force F (measured at the same loading rate) has the same value for both studied molecular complexes but the obtained relations between the unbinding force and loading rate intersect (Fig. 5.9c). Two different slopes denote two different widths of the energy barrier (x_{b1} and x_{b2}), the energy barrier height of the *complex_2* is larger than for *complex_1* (from intercepts comparison) and since the unbinding force F_1 is larger, the applied external force F_1 reduces more the energy barrier height for *complex_1* compared with *complex_2* ($F\, x_{b1} > F\, x_{b2}$).

5.1.7 Other Theoretical Models for Single Molecule Interactions

The Bell–Evans approach is the oldest and most widely applied model describing the unbinding of single molecular complexes [27]. However, the data collected so far show that for certain cases this model does not provide satisfactory description of the unbinding of single molecular complexes. The more realistic models like the Dudko–Hummer–Szabo [28, 29] or the Friddle–Noy–De Yoreo [30] ones have been developed.

5.1.7.1 Dudko–Hummer–Szabo model

In the Dudko–Hummer–Szabo model [28, 29], the single molecular complexes unbinding is described by a harmonic free-energy potential with a single sharp energy barrier. The molecular complex is pulled apart at constant velocity by the external force representing a harmonic spring. The relations for the most probable unbinding force and the probability distributions are very similar to those obtained on a basis of the Bell–Evans model. However, to describe the unbinding process more efficiently, an additional parameter, i.e., energy barrier height, has been introduced in the equation for the unbinding rate:

$$k_{\mathrm{off}}(F) = k_0 \cdot \left(1 - \frac{a \cdot F \cdot x_{\mathrm{b}}}{E_{\mathrm{b}}}\right)^{\frac{1}{a}-1} \cdot e^{\left[1 - \left(1 - \frac{a \cdot F \cdot x_{\mathrm{b}}}{E_{\mathrm{b}}}\right)^{\frac{1}{a}}\right]}, \tag{5.33}$$

where a corresponds to the shape of the free-energy potential. It is equal either to 3/2 or to 1/2, i.e., assumes either linear-cubic or cusp-like shape of the energy barrier, respectively.

In the Dudko–Hummer–Szabo model, the relation between the most probable unbinding force and the loading rate is the following:

$$F_{\mathrm{unb}}(r_{\mathrm{f}}) = \frac{E_{\mathrm{b}}}{a \cdot x_{\mathrm{b}}} \cdot \left[1 - \left[\frac{k_{\mathrm{B}} \cdot T}{E_{\mathrm{b}}} \cdot \ln \frac{k_0 \cdot k_{\mathrm{B}} \cdot T \cdot e^{\left(\frac{E_{\mathrm{b}}}{k_{\mathrm{B}} \cdot T} + \gamma\right)}}{r_{\mathrm{f}} \cdot x_{\mathrm{b}}}\right]^{a}\right], \tag{5.34}$$

where $\gamma = 0.5772$ is the Euler–Mascheroni constant.

The introduction of the energy barrier height E_b allows to interpret cases with multiple energy barriers characterized by widths x_b of less than 1 Å such as for avidin-biotin [15] or individual Fv fragments of anti-lysozyme antibodies [31].

5.1.7.2 Friddle–Noy–De Yoreo model

In the Friddle–Noy–De Yoreo model, when two molecules are pulled apart, two cases are considered, namely, (i) an equilibrium one at lower loading rates that enables for rebinding of molecules and (ii) a so-called kinetic phase observed at higher loading rates where molecules can unbind reversibly. The most probable unbinding force is approximated by the equation

$$F_{unb}(r_f) = f_{eq} + \frac{k_B T}{x_b} \cdot \ln\left(1 + \frac{x_b \cdot r_f \cdot e^{-\gamma}}{k_0(f_{eq}) \cdot k_B \cdot T}\right), \quad (5.35)$$

where γ = 0.5772 is the Euler–Mascheroni constant, f_{eq} is the force value at which dissociation and association rates are equaled to

$$f_{eq} = \sqrt{2 \cdot k_c \cdot E_b}, \quad (5.36)$$

where k_c is the cantilever spring constant.

5.2 AFM Measurements of Adhesive Properties

In the experiments carried out using such techniques as surface plasmon resonance [20] or quartz crystal microbalance (QCM) [31], the binding/unbinding properties are deduced from the measurements where large number of molecules participate and it is not always easy to deduce the kinetics at a single molecule level. Studies of single molecule processes require techniques characterized by high spatial and temporal resolution. Such methods encompass mainly a biomembrane force probe with pipette suction [32], a hydrodynamic flow-based method [33], magnetic [34] and optical tweezers [35], and also atomic force microscopy (AFM) [36–38].

The most attractive feature of the AFM technique is the capability to quantify the unbinding forces with the participation of receptors/ligands present directly on a surface of living cells. In such measurements, both the AFM probe and studied cells are immersed in a buffer imitating the native conditions. The measurement itself can be realized in two ways. First, the cantilever surface is coated with ligand molecules, then such modified AFM probe is brought into contact with the cell's surface (Fig. 5.10).

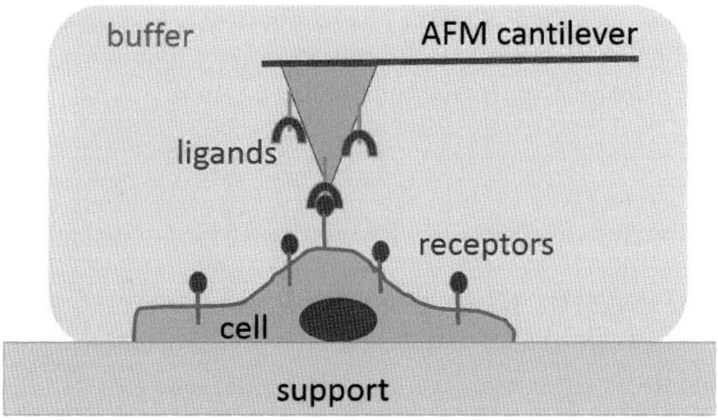

Figure 5.10 The schematic illustration of the measurements of cellular adhesive properties in living cells, carried out by atomic force microscope using a ligand modified cantilever interacting with receptors presented on a cell surface.

The ligands present on a surface of AFM cantilever recognize specific types of cell surface receptors. The other approach uses a single cell as probe (Fig. 5.11). If the substrate surface is coated with ligands specific to certain type of cellular receptors, the measurements deliver the information similar to the case when ligand-modified cantilever probes the cell surface.

When cells are cultured on a support, the measured forces reveal the cell-to-cell adhesion interactions. Independently of the applied approach, the analysis of adhesive properties in living cells uses the retract part of the force curve recorded during the AFM cantilever withdrawal. Analogously to elasticity measurements, several aspects, specific to unbinding measurements, should be considered to obtain reliable information on magnitude of the unbinding forces.

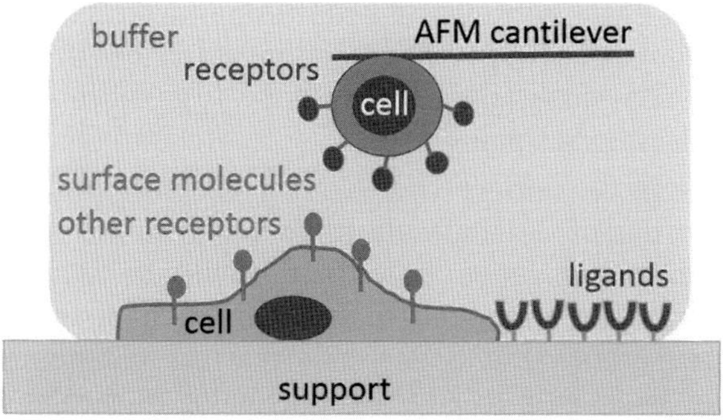

Figure 5.11 The schematic illustration of the measurements of cellular adhesive properties, using a cell probe that can be brought in contact either with a surface modified with isolated ligands or with a surface of another cell.

5.2.1 Attachment of Molecules to Desired Surfaces

Studies of the adhesive properties in living cells require the application of protocols that are used to modify surfaces of both cantilevers and substrates. The group of molecules specifically recognized by the cell surface receptors includes almost all molecules present in nature. The molecules of interest can be broadly grouped into (1) nucleic acids (DNA, RNA), (2) proteins (antibodies, enzymes, and receptors), (3) small molecules (e.g., peptides, metabolites), and (4) other biomolecules, like carbohydrates or lipids. Their attachment to various surfaces depends strongly on their surface properties, on the chemistry of substrate, and/or on the liquid medium composition.

5.2.1.1 AFM probe functionalization

Such molecules like proteins exhibit different structural properties, manifesting in highly heterogeneous hydrophobicity, and charge distribution. This can make their deposition on the AFM probe very difficult. In particular, for all of them, it is essential to preserve biological activity during the experiment. Additional complications can arise when a correct orientation of the bound molecule is required. The immobilization protocols used to attach molecules employ both adsorption and covalent binding. Although many various protocols have been developed, several basic issues always have to be considered:

(1) Choice of an appropriate AFM probe in terms of a spring constant value (due to detection limit described in Chapter 3); a tip sharpness, surface chemistry (e.g., silicon or silicon nitride or gold coating).
(2) Knowing the surface chemistry of the AFM probe and the properties of the molecule to be attached, the corresponding functionalization procedure should be selected. It is important to remember that during force-induced unbinding experiments, it is desirable that the unbinding should occur between a pair of two interacting molecules that are studied (the strength of the molecule attachment to the tip surface should be larger than the interaction between the studied molecules);
(3) The density of molecules on the surface of the AFM probe should be considered, since low concentrations can not only reduce multiple interactions but also eliminate the occurrence of single unbinding events.
(4) The decision whether to use or not polymeric spacers should be taken after considerations whether spacer may help to recognize specific unbinding events.
(5) During the AFM probe functionalization (and also measurements), environmental factors such as buffer compositions, pH or temperature should be maintained to assure binding activity unchanged.
(6) When molecules are needed to be attached on a support like mica or glass surface, their roughness should be smaller than the diameter of molecules.

The choice of chemical reagents depends strongly on binding targets (several examples are provided in Table 5.1).

Table 5.1 Common binding targets used in the AFM probe functionalization

Target functional group	Found in	Reactive group	Bond type
–COOH (carboxyl)	Aspartate (AA) Glutamate (AA)	NH_2 OH	Amide Ester
$-HN_2$ (amine)	Lysine (AA) Silane treated surface Ethanolamine treated surface	NHS-ester Carboxyl	Amide Ester

Target functional group	Found in	Reactive group	Bond type
–SH (sulfhydryl)	Cysteine (AA) Thiol	Maleimide carboxyl	Thio-ester Thio-ether
–CHO (carbonyl)	Oxidized carbohydrates	Hydrazide	Hydrazone
–OH (hydroxyl)	Serine (AA) Theronine (AA)	Carboxyl	ester
Avidin	Avidin modified proteins	Biotin	Avidin-Biotin bond

AA: amino acid.

Actually, there are two popular approaches that either directly or indirectly attach molecules to the surface of the AFM tip (Fig. 5.12).

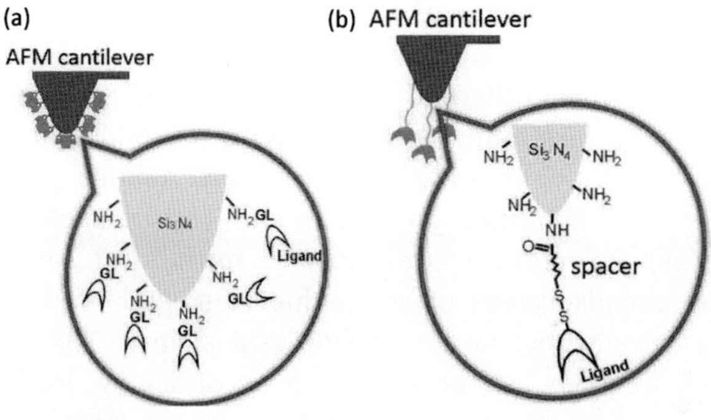

Figure 5.12 Two ways of the AFM tip functionalization. The desired molecules are either (a) directly attached to the probe surface through cross-linking agents as glutaraldehyde (GL), or (b) indirectly through a polymeric spacer (e.g., PEG).

The first step in the approach enabling the direct immobilization of molecules to the AFM probe, is the surface silanization with 3-amino-propyltriethoxylsilane (APTES[3]), which enriches the surface with amino groups [36]. Then, the silanized surface is activated using a cross-linking agent, such as glutaraldehyde,

[3]When buying APTES, it is very important to ask the AFTES provider for the fresh compound since the water arriving from the air can lead to deactivate the formation of amino groups on the cantilever surface. The solution should be stored in the fridge with a cap wrapped with parafilm.

which binds a protein through amine group (Fig. 5.12a). Such a way of AFM tip functionalization results in a randomly attached protein molecules without knowing and controlling their density and orientation. Thus, only a small percentage of molecules will be oriented in such a way that they can interact with the corresponding molecule (receptor) present on a cell surface. The main advantages of this approach are the simplicity and generality. The AFM probes can be functionalized during a relatively short time (less than 2–3 h), which provides always freshly prepared AFM probes used during the measurements. The main drawback is a lack of protein flexibility, which leads to low efficiency of measurements and difficulties in identification of the specificity interactions, since very often non-specific forces are difficult to be excluded from the analysis.

To provide better flexibility of the interacting molecules, the polymeric spacers (linkers) can be applied (Fig. 5.12b). The attached spacers provide the molecules the freedom to move around and also prevents their denaturation [39]. Usually, such spacers carry two different functional groups, namely: a NHS group reacting with amine groups present on AFM probe surface and PDP (2-pyridyldithiopropionyl) or vinyl sulfone groups that covalently bind to thiols present in the desired molecule (e.g., in proteins). Polymeric spacers can be introduced, for example, either by chemisorption of alkanethiols on gold or by covalent coupling of polyethylene glycol (PEG) to silanes. The degree of molecules freedom can be enhanced by attaching, for example, recombinant histidine-tagged proteins onto an AFM tip coated with nitrilotriacetate (NTA)-terminated alkanethiols [40]. The optimal length of PEG spacers is still under debate [41–43]. Thus, it should be adjusted to molecular complex studied. It has been demonstrated that very long chains (above 35 nm [42]) can decrease significantly the binding probability and, simultaneously, the efficiency of the measurements.

The molecules immobilization onto specific surfaces like mica or glass involves described two major categories of mechanisms: adsorption and covalent binding.

Adsorption relies on non-covalent interactions—mainly electrostatic, van der Waals, and dehydration of hydrophobic interfaces [43]. It has a purely physical nature and therefore displays varying levels of reversibility. The adsorption of proteins depends on two main features: their surface charge and their hydrophobic domains. Both properties enable a certain control of protein deposition; however, they can result in randomly oriented molecules. The electrostatic adsorption seems to be sufficient to assure a relatively strong attachment, but it does not have permanent nature and it can be strongly affected by changes of solution pH and ionic strength. Therefore, only a limited number of proteins can be immobilized in this manner. When the hydrophobic attraction is chosen as a main source of adsorption, stronger and less reversible interaction is expected. On the other hand, it may result in loss of functional activity due to partial denaturation, as the protein unfolds to expose hydrophobic interior portion to the hydrophobic surface. Such way of molecule immobilization results in their random orientation and relatively weak attachment, which may significantly elongate the time of measurement.

The covalent binding of specific molecule functional groups to functionalized surfaces, by definition involves formation of essentially irreversible chemical bonds between the molecule and the substrate surface. However, in many cases the covalent binding is enabled only after additional functionalization of the surface and/or biomolecule [44, 45]. A variety of side groups can be easily used for covalent binding—most common ones are amino, carboxylic, hydroxyl, and thiol groups. Thus, this resulted in many strategies for cross-linking of available functional groups. Most of them use specific cross-linkers, for both attachment and physical separation of protein from the surface, thereby allowing larger fraction of the protein functional domains to be exposed to the buffer. Covalent binding generally produces a higher concentration of proteins than the adsorption. Proteins can be also better oriented by additional techniques, such as

(1) use of antibodies that bind proteins leaving the antibody binding sites free;
(2) use of biotinylation which enriches proteins with the binding site specific to streptavidin-coated surface;
(3) cysteine thiol production in the protein fragment far from the binding site, allowing its deposition on gold coated surface;
(4) use the sugar molecules that bind to the oligosaccharide's moieties of proteins.

In biological applications, the most common AFM probes are made of silicon or silicon nitride. Biomolecules are usually immobilized on glass, mica, and gold. Therefore, immobilization requires a development of an appropriate protocol of the attachment. Such way of molecules immobilization results in a very strong attachment and, in certain instances, enables the oriented molecule deposition.

5.2.1.2 Preparation of a cell probe

The use of a single alive cell as an AFM probe requires to attach it at a very end of the cantilever. In most cases, tipless cantilevers are applied. Several approaches have been already published. In one of them, cells are cultured directly on the cantilever surface for a couple of hours before the AFM experiment. Then, only cantilevers with a single cell attached at the end of the cantilever can be chosen for measurement [37]. The advantage of this approach is a strong attachment of a cell to the cantilever surface; however, the drawback is a waste of high number of cantilevers that are not suitable for the experiments.

The other protocol for a single cell attachment uses the ideology similar to the functionalization of the cantilever surface with single molecules; however, the choice of the strategy depends on the cell dimension and its surface properties. Figure 5.13 presents graphically the most popular protocol that involves the functionalization of the cantilever surface with a lectin—Concanavalin A. It follows the protocol previously derived by Wojcikiewicz et al. [44], and it is schematically shown in Fig. 5.13a. Concanavalin A (Con A) is a glycoprotein derived from *Canavalia ensiformis* that specifically binds to mannose or glucose residues of oligosaccharide moieties covalently attached to glycoproteins or glycolipids, present on a cell surface [44, 46]. In the easiest and also less time-consuming protocol, a cantilever surface is activated with oxygen plasma (simultaneously, oxygen plasma cleans the surface). After plasma activation, cantilevers are immersed in Con A solution, followed by washing them in the buffer (Fig. 5.13a).

Cells, previously cultured in a plastic dish are trypsinized, dissolved to obtain their low concentration (too high number of cells would results in more than one cell attached to the cantilever), and placed over a glass slide in a buffer used for the AFM measurements (Fig. 5.13b). Con A modified cantilever is placed over a single cell and gently approached towards the cell. Then, a cell is delicately indented for about 5 s. Next, the cantilever is retracted until a cell detaches from the glass surface (the retraction distance should be around 100–200 microns to prevent the cell to re-attach to the glass surface). After 15–20 min, the cell is usually sufficiently strongly attached to the cantilever and it can be used in the AFM measurements (Fig. 5.13c). The perfect situation is when a single cell is placed in a central part at the end of the cantilever. If not, one can try a very delicate scan to roll a cell towards the cantilever center.

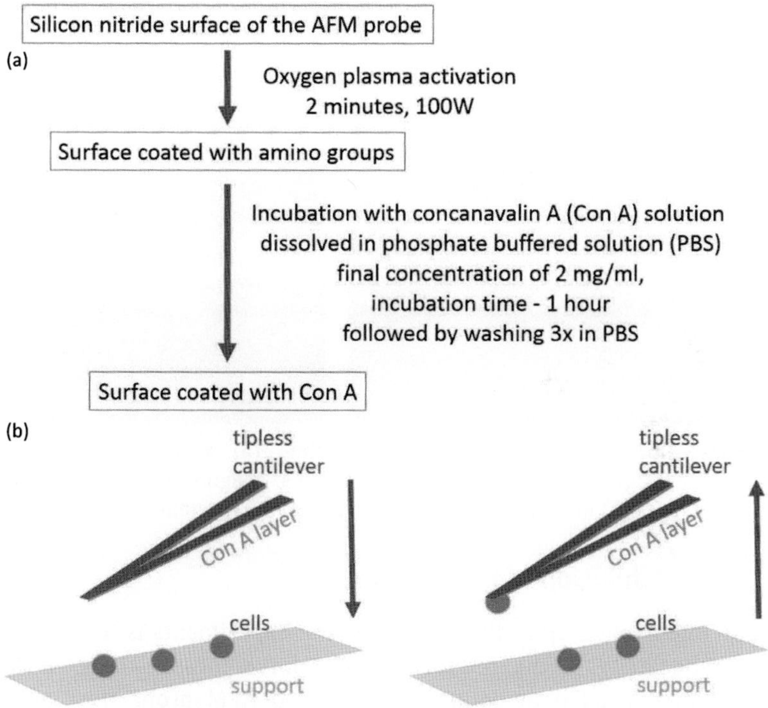

Figure 5.13 (a) Exemplary protocol used to functionalize the silicon nitride surface with concanavalin A. (b) The idea of the single cell catching in buffer solution with rounded cells, added directly after their trypsinization. Inset: the image of single melanoma on silicon nitride cantilever recorded by inverted optical microscope (tipless, NP-O, Brucker).

5.2.1.3 Cells preparation for the AFM measurements

The use of the AFM for experiments with living cells requires them to be well-attached to the substrate surface. Usually, cells are cultured either on a glass coverslip or at the bottom of a plastic Petri dish. In most cases, no additional preparation such as paraformaldehyde fixation, etc. is needed, since cells are rather strongly attached to the surface if they are of adherent type and culture conditions are kept properly. If cells are weakly attached, the special adhesives such as poly-L-lysine, fibronectin, collagen, laminin, or CellTak™ can be used to enhance the cellular adhesion.

The density of cells depends on the aim of measurements. In most cases, single cells are measured. Figure 5.14 shows typical density of cells used in the AFM measurements (images recorded by (a) top view optics and (b) inverted optical microscope).

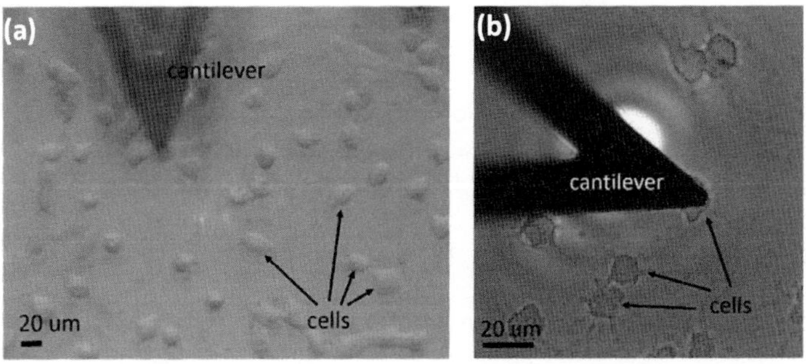

Figure 5.14 Images of cells recorded using: (a) a top view optics (AFM type Xe120, Park Systems, magnification 20×) and (b) inverted microscope (Olympus IX9, magnification 40×).

5.2.2 Inhibition of Binding Site

The important step in the AFM adhesion experiments is to assure the specificity of the interaction. The force measured by AFM is a force needed to detach the modified AFM probe from the investigated surface. Therefore, additional experiments proving that the observed interactions reflect the specific recognition between single molecules must be carried out. One way is to use standard biochemical methods such as fluorescence microscopy [47], immunolabeling [48], quartz micro-balance [49], etc. However, very often these methods are not sensitive enough to prove the specificity of the interaction. Thus, apart from them, the common way of carrying control measurements is to block the studied interaction by adding to the solution the same type of molecules as those attached to the AFM probe surface (Fig. 5.15a).

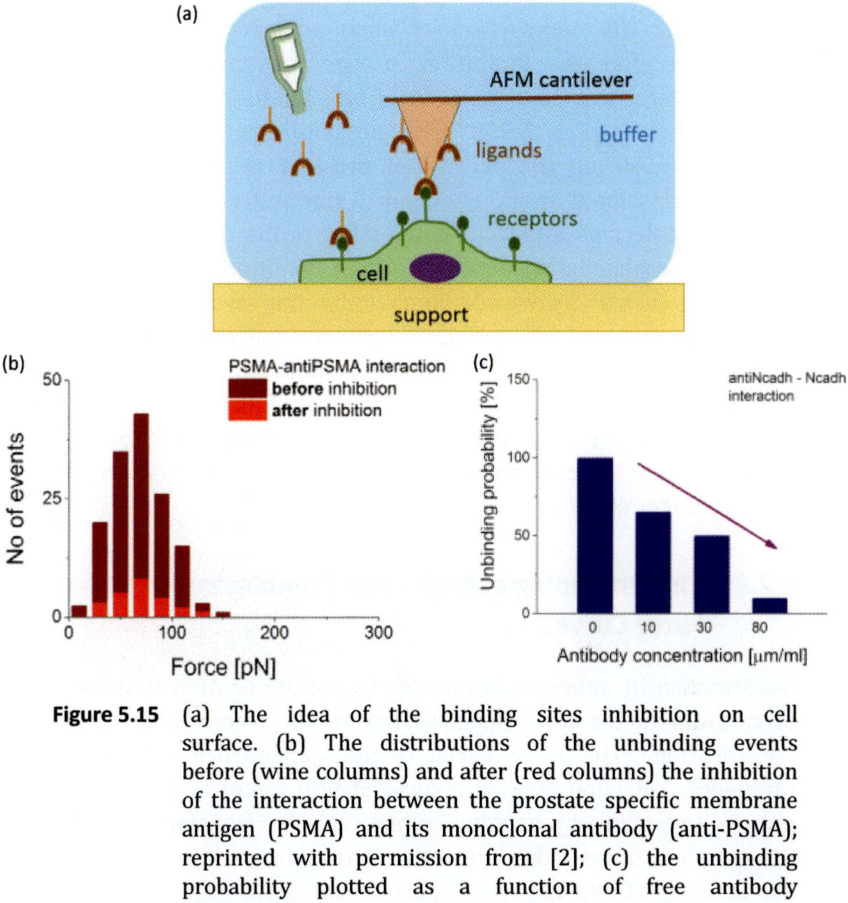

Figure 5.15 (a) The idea of the binding sites inhibition on cell surface. (b) The distributions of the unbinding events before (wine columns) and after (red columns) the inhibition of the interaction between the prostate specific membrane antigen (PSMA) and its monoclonal antibody (anti-PSMA); reprinted with permission from [2]; (c) the unbinding probability plotted as a function of free antibody concentration—an example of the interaction measured between monoclonal antibody against N-cadherin and N-cadherin in T24 bladder cells.

These free molecules bind to the binding sites of the corresponding receptors present on a cell surface and make them inactive during measurements with the functionalized AFM probe. The inhibition experiments can be parameterized by *the unbinding probability*, which is defined as a ratio between the number of unbinding events to the total number of force curve recorded. This quantity is proportional to the number of binding sites of receptor molecules (or to the receptor number if only one binding site is present on its surface). The inhibition of binding sites results in the lower number of events (and thus smaller event probability). The comparison of such estimated probabilities, before and after the inhibition, proves the specificity of the studied interaction. The example of the binding site inhibition is presented in Fig. 5.15b. The prostate specific membrane antigen, present on the surface of prostate cells, was blocked with its monoclonal antibody added to the buffer [50]. Thus, the antibody molecules attached to the AFM probe did not interact with the complementary antigen and the resulting number of the unbinding events decreased. Some unbinding events were still observed, since not all antigens were recognized and blocked by the antibody molecules. Another example (Fig. 5.15c) shows the decrease of the unbinding probability as a function of antibody concentration (for the interaction studied between monoclonal antibody against N-cadherin and N-cadherins present on a surface of T24 bladder cells).

5.2.3 The Unbinding of Molecular Complexes: Force Curves

Qualification of adhesive properties by means of AFM is usually carried out in the force spectroscopy mode, where *force curves* are recorded. The force curve is the dependence between a cantilever deflection (that is converted into force) and a relative sample (or scanner) position, which can be transformed to a tip–sample distance. The interaction forces can be obtained from the analysis of the retraction part of the force curve recorded during the AFM cantilever withdrawal from the surface.

In cases, when individual complexes are ruptured with the participation of cell surface, the unbinding can proceed through three most common scenarios. Ideally, only the rupture of a single complex occurs (Fig. 5.16). Then, the resulting force curve contains the characteristic *"jump"* that can be parameterized by two quantities: the unbinding force F and the rupture length L.

When two (or more) single complexes of the same type are simultaneously ruptured (Fig. 5.17a), the character of the force curves changes, depending on how many complexes are formed within the contact area between the ligand-modified tip and the receptor-covered surface and what the mechanism of the multiple unbinding is (two simplest cases are cooperative and uncooperative one). For cooperative unbinding, multiple molecular complexes break simultaneously, which manifests in the larger unbinding force (being the product of a force F obtained for the single complex and the number of complexes n ruptured at the same time) and in the similar length L as observed for the single molecular complex. The resulting force curve has a similar shape with larger unbinding force (Fig. 5.17b). In such case, the length histogram would have only one peak at the most probable rupture length L but in the force histogram, multiple maxima are expected at positions of F, $2F$, ..., nF values. When molecular complexes act uncooperatively, the rupture proceeds sequentially and the detachment force has a similar value as that of a single molecular complex. The force curve will show a saw-tooth pattern with unbinding events of the similar rupture length L and the unbinding force F values (Fig. 5.17c).

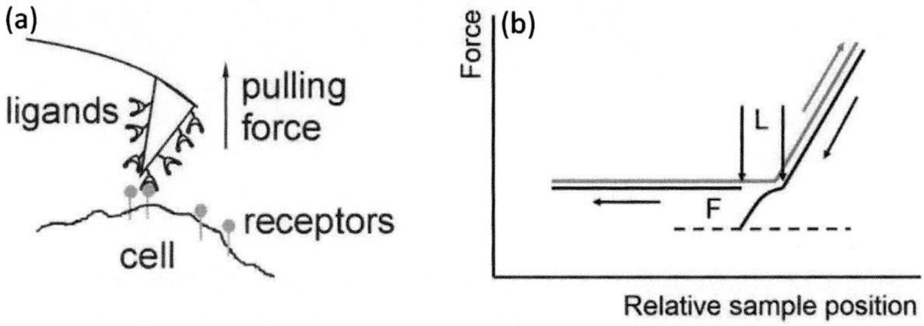

Figure 5.16 (a) Illustration of a single molecular complex unbinding. (b) An ideal force curve characteristic for the interaction between a single pair of ligand and receptor molecules (red line—approach part, black line—cantilever withdrawing denotes the curve recorded during molecular complex unbinding). Reprinted with permission from [2].

These considered possible shapes of force curves are also valid for experiments with the use of isolated proteins where one protein is attached to the AFM probe and the other, complementary one, is immobilized on the substrate surface. The embedding of receptors in a plasma membrane can often influence the unbinding process by the induction of the membrane deformation that may manifest in the broadening of the rupture length range but the unbinding force remains unchanged (unless any unrecognized non-specific interaction is present in the analyzed data). Thus, the unbinding events obtained for a given receptor, embedded in the cell membrane, produce the maximum in the force histogram and a broad distribution of the rupture lengths with or without distinguished peaks. In this case, the mean value of the rupture length will be shifted to higher values.

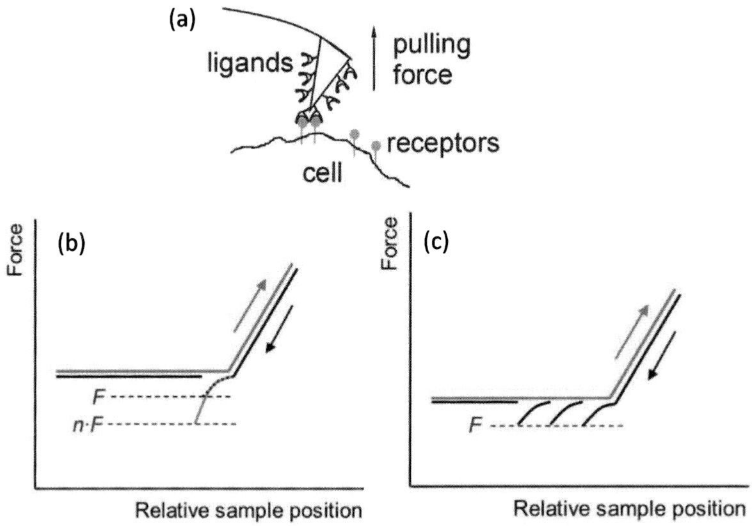

Figure 5.17 Illustration of simultaneous unbinding of two molecular complexes (a) showing characteristic force curves for (b) cooperative and (c) uncooperative unbinding (red line—approach, black line—unbinding curves. Reprinted with permission from [2].

5.2.4 Parameters Derived from a Single Force Curve

To quantitatively characterize the specific interaction between a pair of molecules, a set of distinct parameters is derived from the force curves. An additional parameter, determined on the basis of the total number of performed measurements, is an unbinding probability that is related to the number of receptors present on the substrate surface or in the cell membrane.

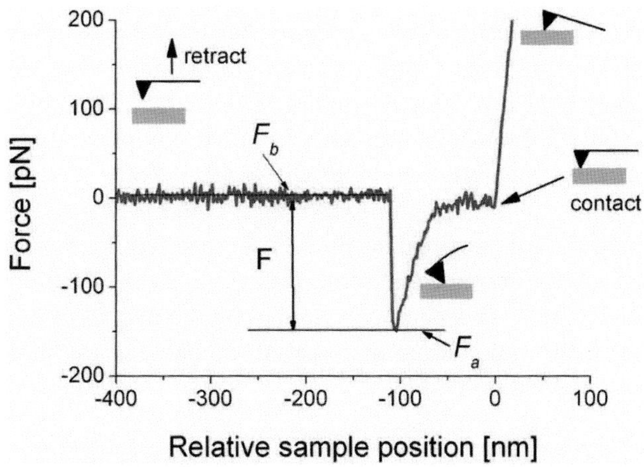

Figure 5.18 Example of the force curve recorded for the interaction of an antigen–antibody pair, showing a typical jump related to single molecule interaction. Reprinted with permission from [2].

5.2.4.1 The pull-off force and force histogram

The unbinding force needed to separate two interacting molecules is delivered from the analysis of the retraction part of the force curve. It should be pointed out here that the AFM can measure only a so-called pull-off force (Fig. 5.19).

The pull-off force (called also rupture or detachment force) is defined as a difference between the force F_b, corresponding to the free cantilever position (when the interacting force is negligible) and the maximum value of the force F_a:

$$F = |F_\mathrm{a} - F_\mathrm{b}| \tag{5.37}$$

The pull-off force value measured for two surfaces, bearing the complementary proteins that interact in a specific way, is usually a superposition of two components: (i) discrete, short-range one, dominating within the binding sites that is related to the strength of a single molecular complex (*referred here as specific forces*), and (ii) that originating from long-range, distance-dependent forces dominating outside of the binding site (*referred here as non-specific forces*). The non-specific forces vary in response to the properties of the environment surrounding both molecules of interest. Very often, it is difficult to separate between the specific and non-specific interactions since the strength of the latter one can be comparable. For example, the determined non-specific forces [51] ranged from 60 pN to about 400 pN (!), which was significant in comparison with the measured specific interaction forces (240 ± 160 pN for ConA–ASA, 370 ± 110 pN for ConA–CaY and 180 ± 130 pN for PAP–aPAP pairs). For analogous protein embedded in the plasma membrane, the non-specific force was 60 ± 30 pN for ConA-PC3 cells. Such a low value can be explained by the overall interactions present on the cell surface that are not involved in the molecular recognition phenomenon.

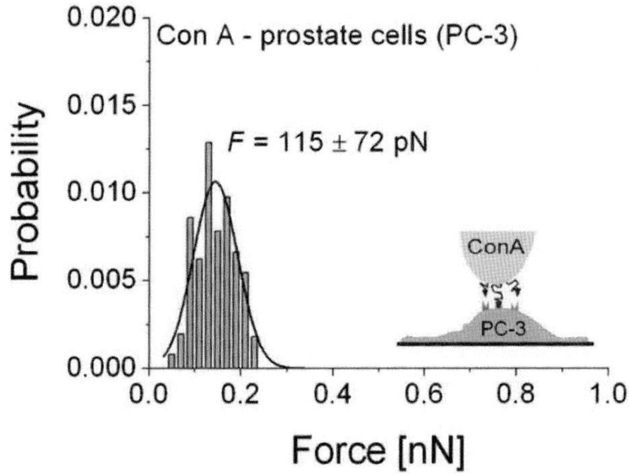

Figure 5.19 Force histograms showing only one peak attributed to the specific interaction between mannose-type glycans present on the surface of prostate (PC-3) cells probed with AFM tip functionalized with lectin ConA. The solid line is a Gaussian fit used for determination of the unbinding force. Reprinted with permission from [2].

The force histogram is created using the bin size reflecting the minimum value of detected forces (see Chapter 3 for more details). Shape of the histogram is characteristic for the studied pair of molecules and depends on the number of molecules present within the contact area. To quantify the unbinding force, many force curves must be collected and analyzed using histograms that can reveal either only one peak (Fig. 5.19) or multiple peaks (Fig. 5.20).

The single force maximum can be directly attributed to a rupture of a single molecular complex unbinding. The multiple peaks can correspond to the case of simultaneous rupture of more complexes (Fig. 5.20).

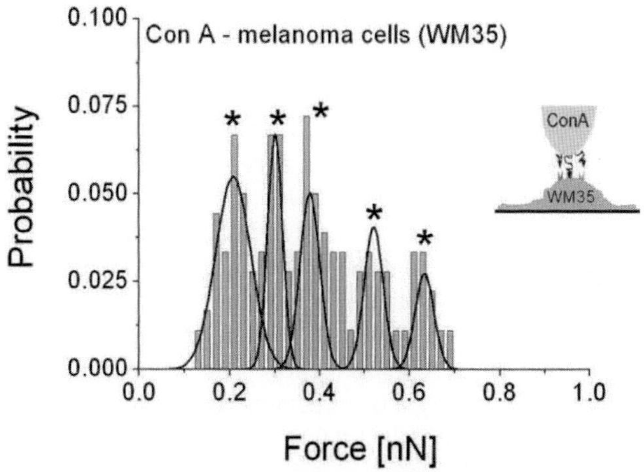

Figure 5.20 Force histograms showing multiple peaks observed when surface of melanoma cells (WM35) is probed with ConA-modified tip. The solid line is a Gaussian fit used for determination of the unbinding force. Reprinted with permission from [2].

The most probable unbinding force is usually determined by fitting the Gaussian functions to the maxima present in the histogram of the measured rupture force. Unbinding force is calculated as a center of the fitted Gauss distribution and the corresponding error is a standard deviation determined from the half width of the peak at its half maximum height. Such fitting procedure gives the position of the maximum with reasonable reliability but it does not describe fully the shape of force histogram which is determined by the stochastic nature of the unbinding process. It can be analytically described by the probability density function:

$$y(x) = a \cdot e^{b \cdot x} \cdot e^{-c \cdot (e^{b \cdot x})}, \tag{5.38}$$

where a, b, and c are parameters fitted to the unbinding force histogram obtained at the given loading rate r_f. The probability density function is determined by the two parameters describing rupture of a single molecular complex: the energy barrier x_b and the dissociation rate constant k_0. The fit delivers the estimates of those parameters. They are calculated according to

$$x_b = b \cdot k_B \cdot T \tag{5.39}$$

$$k_0 = \frac{x_b \cdot r_f \cdot c}{k_B \cdot T} \tag{5.40}$$

The dissociation rate constant and the energy barrier position determined in such way were in a reasonable agreement with values were obtained using two independent experiments [46, 48], namely, using the QCM and the AFM working in a dynamic force spectroscopy mode (DSF). The dissociation rate constant of 0.137 ± 0.029 s^{-1} obtained from the fit is placed between values 0.095 ± 0.002 s^{-1} (from QCM measurements) and 0.170 ± 0.060 s^{-1} (from AFM measurements). The fitted position of the energy barrier was in good agreement with the reported value (0.23 ± 0.01 nm versus 0.229 ± 0.004 nm).

5.2.4.2 Relation between the unbinding force and the number of ruptured bonds

When multiple peaks are observed in a force histogram, they are usually attributed to the rupture of more than one single complex present within the contact area of the AFM probe and the surface. Thus, the first peak corresponds to the unbinding event involving the rupture of one single complex; the second one is related to the simultaneous unbinding of two single complexes (thus, the force value at the second maximum is doubled), etc. Such force histogram can be translated into the relationship of the unbinding force determined for each consecutive peak and the peak number (i.e., number of ruptured single complexes), and the linear dependence is expected if only one type of interaction is present (Fig. 5.21). The peak number is equal to the number of ruptured single complexes when the center of the first force peak agrees with the force value obtained from the slope. The unbinding force for a given individual lectin–glycan complex, i.e., ConA–mannose–type glycans, determined from the slope of the fitted line, was of 105 ± 15 pN. This value is in agreement with those obtained to unbinding ConA from the same glycan type probed on the prostate cells (115 ± 72 pN) [50]. Therefore, the position of the first peak (~200 pN) can be attributed to the simultaneous rupture of two single complexes, which was confirmed by the large number of mannose-type glycans.

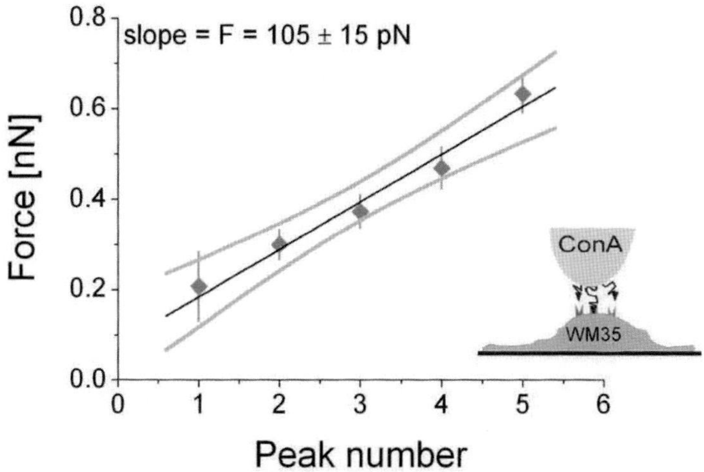

Figure 5.21 Linear regression fitted to the unbinding force as a function of the number of succeeding peaks observed in the histogram (Fig. 5.20b) obtained for melanoma WM35 cells probed with lectin ConA. Data points correspond to centers of Gaussians fitted to each single peak present in force histogram while error bars represent their standard deviations. The 95% confidence bands are marked as grey lines. Reprinted with permission from [2].

In order to statistically evaluate the obtained differences between cell lines, for each fitted line the confidence bands of 95% were calculated. They estimate the certainty of the shape of the fitted line and the assumed confidence level implies a 95% chance that the true regression line fits within these bands (grey lines in Fig. 5.21). This approach works reasonably well when the regression curve is calculated basing on more points, which in our studies corresponds to cases where histograms were composed of five maxima.

5.2.4.3 The rupture length and its histogram

If the single complex is formed, the AFM probe withdrawal generates bending of both interacting molecules. Then, the rupture length can be defined an unbinding distance L^* calculated after the conversion of the force curve into the force versus tip–surface distance function (see Chapter 3 for details). The value of the rupture length, L, is determined from the following equation:

$$L = L^* - \frac{F}{k_c} = |L_k - L_c| - \frac{F}{k_c}, \qquad (5.41)$$

where k_c is the cantilever spring constant, F is the unbinding force, the distance $|L_k - L_c|$ is the distance between the moment when the cantilever starts to bend (L_c) and the moment when molecules unbind (L_k) as it is shown at Fig. 5.22.

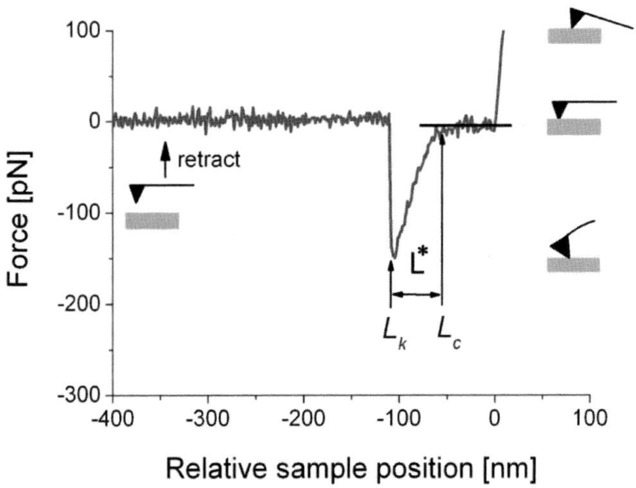

Figure 5.22 Example of the force curve recorded for the interaction between antigen–antibody complex. Reprinted with permission from [2].

The value of the rupture length L brings information about the mechanical resistance of the formed complex to the applied external force. The exemplary distribution of the rupture length obtained for the unbinding of single anti-BSA-BSA complexes is presented in Fig. 5.23.

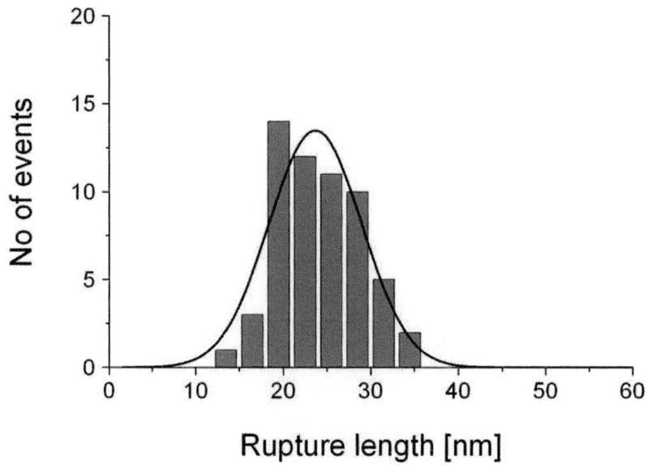

Figure 5.23 Histogram of the rupture length obtained for the interaction between albumin molecules and monoclonal antibody against albumin. The monoclonal antibody against albumin (anti-BSA) was immobilized on a surface of the AFM probe while albumin molecules (BSA) were deposited on a mica surface. The line denotes the Gauss function. Reprinted with permission from [2].

The distribution shows only one peak centered at 23.7 ± 5.3 nm (the bin size reflects the experimental detection limits—the distance between two subsequent z-steps was 3 nm) and they agree with the data reported for BSA and polyclonal antibody against BSA [52]. However, if the single molecular complex is formed between a ligand and a receptor embedded in the cell membrane, the rupture length would reflect the overall mechanical resistance of the whole system composed of these two molecules and the cell membrane at the place of the receptor anchorage. Since most receptors present on a surface of living cells are linked with actin cytoskeleton, the mechanical resistance will be also influenced by properties and structure of such a linkage.

5.2.4.4 The number of ruptured bonds

Very often, during the measurements on the surface of living cells, force histograms show multiple maxima (like for example in Fig. 5.20b) that correspond to the simultaneous rupture of subsequent complexes. To quantify this case, the area under each peak can be used as an indicator of the probability of simultaneous rupture of *n* bonds.

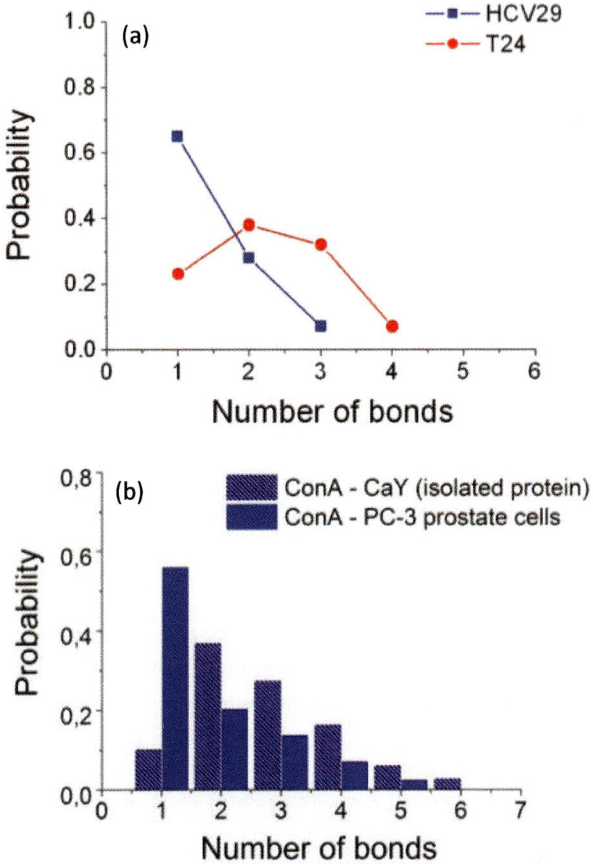

Figure 5.24 Probability of the simultaneous rupture of *n* bonds obtained for (a) N-cadherin–antibody complex in cancerous T24 and reference HCV29 cells, (b) for the lectin concanavalin A (ConA) and glycans either attached to isolated protein (carboxypeptidase Y, CaY) or present in a plasma membrane of prostate cells (cell line: PC-3). The distributions were normalized to the total number of events.

The probability of the simultaneous rupture of *n*-bonds, obtained for the same interaction occurring between N-cadherin and its monoclonal antibody in cancerous T24 and reference HCV29 cells, is shown in Fig. 5.24a. One can observe that the simultaneous rupture of 2–3 bonds was the most probable for the cancerous cells (T24) as compared to the reference cells (HCV29) where the rupture of a single bond was most prominent. When lectin was used as a probe for mannose bearing glycans, simultaneous rupture was observed when these glycans were attached to the isolated protein where they were better accessible (Fig. 5.24b). Such situation indicates larger number of active, isolated and single protein molecules on the investigated surface.

5.2.4.5 The unbinding probability

Another quantity related to the number of molecule present on the surface is the unbinding probability $P_{unbinding}$ that is usually defined as a ratio between the number of curves with the unbinding events to the total number of the recorded force curves:

$$P_{unbinding} = \frac{\text{Number of curves showing unbinding events}}{\text{Total number of the recorded force curves}} \tag{5.42}$$

Its value corresponds directly to the number of molecules present on a surface of living cells, which can vary depending on the complex or cell types [53] (see also Table 5.2).

Table 5.2 The unbinding probability determined for non-malignant HCV29 and malignant T24 bladder cells [53]

Cells	Lectin-glycan complex	Number of force curves	Number of unbinding events	Unbinding probability
HCV29	ConA[1]–mannose	3982	438	0.110
HCV29	SNA[2]–sialic acid	4057	215	0.053
T24	ConA–mannose	3983	111	0.028
T24	SNA–sialic acid	4095	344	0.084

[1]ConA–lectin from *Canavalia ensiformis*.
[2]SNA–lectin from *Sambucus nigra*.

The determination of the unbinding probability gives only the robust estimation of the number of receptors present on the investigated cells' surfaces. This parameter does not describe the receptor distribution over the cell surface. The same unbinding probability value can be obtained for receptors randomly distributed over a certain area as well as in the case when receptors are grouped around one place in the same area.

5.3 Single Molecule Interaction in Living Cells: A Case Study

The section presents the AFM-based studies on properties of single molecular complexes composed of an individual N-cadherin molecule and monoclonal antibody against N-cadherins (referred here as *Ncadh–GC4 complex*). The force spectroscopy measurement were carried out directly on a surface of living human bladder cells (HCV29—non-malignant cancer cell of ureter and T24 bladder cells from transitional cell carcinoma). The single-molecule measurements were performed with the aim to answer the question whether an individual molecule or its part displays different binding properties in normal and cancerous cells [54]. In these experiments, the monoclonal antibody against N-cadherin was directly immobilized on a surface of the AFM probe. The results presented in this section illustrates the potential of single molecule unbinding experiments in characterization of the given interaction type.

5.3.1 Properties of N-Cadherin in Bladder Cancer Studied by AFM

Cadherins (see Chapter 2) are transmembrane proteins having both extracellular and cytoplasmic domains (Fig. 5.25). The extracellular domain consists of five cadherins repeats (called ectodomains), each of about 110 amino acids residues. Between two repeats, calcium ions are bound, participating in the formation of calcium–dependent, homophilic bonds. The cytosolic domain of the cadherin is directly associated with β-catenin or/and plakoglobin (γ-catenin) [54, 55]. However, there is one exception: Plakoglobin can associate with both classical cadherins (e.g., E- or N-cadherin) and desmosomal cadherins while γ-catenin associates only with the members of the classical cadherin family. Both γ-catenin and plakoglobin bind to γ-catenin, which links the cadherin/catenin complex to the actin cytoskeleton [56, 57].

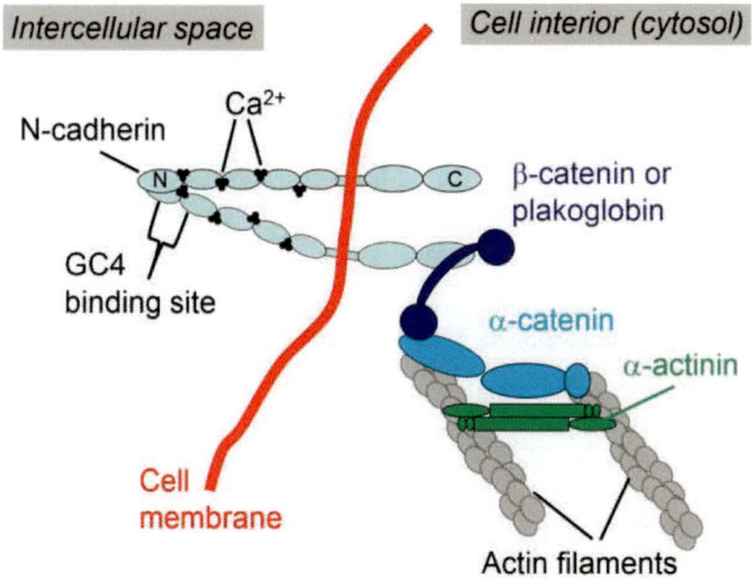

Figure 5.25 Illustration of the cadherin-catenin complex embedded in cell membrane. Reprinted with permission from [2].

In order to detect N-cadherin on a surface of living cells, the monoclonal antibody against the cadherin extracellular domain was applied (GC4 antibody) [58]. The used antibody inhibits adherens junction formation and disrupts existing junctions in cultured cells. The binding site for the antibody is located between the two first, outermost ectodomains ([59], denoted usually as EC1 and EC2, Fig. 5.25).

5.3.1.1 Shape of the force curves for Ncad–GC4 complex

The typical force curves recorded by AFM for the Ncadh–GC4 complex is presented in Fig. 5.26 (only retraction parts are shown).

All analyzed curves showed "jumps" characteristic for the separation of two single molecules. Before the rupture, both molecules were probably stretched. For the Ncadh–GC4 complex studied in both cell types, the force value did not exceed 150 pN but all recorded single unbinding events differed in terms of the unbinding force and the rupture length values. The recorded retrace paths showed jumps characteristic for single molecule unbinding typical for many antibody-antigen complexes [38, 52].

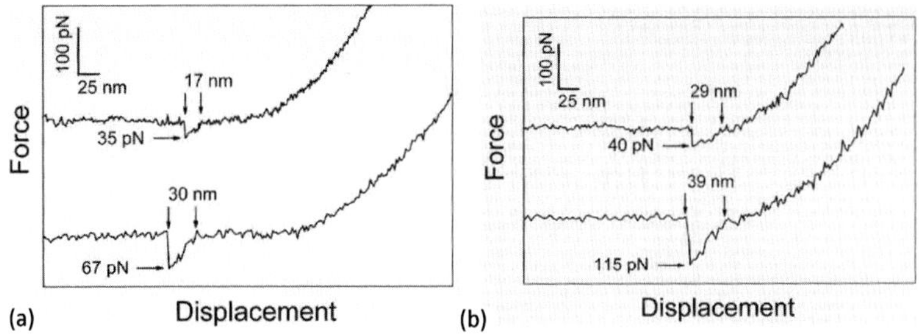

Figure 5.26 Typical force curves recorded for the interaction between Ncad–GC4 complexes measured in both reference HCV29 and cancerous T24 human bladder cells. Reprinted with permission from [54].

Since the binding site for the GC4 antibody is placed in the extracellular domains of N-cadherin, the observed jumps could indicate unfolding of cadherin repeats—ectodomains [58]. In such case, one can observe a characteristic saw-tooth pattern, composed of a few, almost identical jumps corresponding to unfolded single ectodomains, which was not observed during Ncadh–GC4 complex unbinding.

5.3.1.2 Unbinding force dependence on loading rate

The recorded by AFM detachment force corresponds to the most probable unbinding force *F*, whose value is dependent on a loading rate [9]. Thus, for each force curve, the value of loading rate was calculated and the resulted relation for each cell line was drawn (Fig. 5.27).

The loading rate cannot be controlled directly in the AFM experiment, thus its value is derived from the product of the retraction velocity (this parameter can be set in AFM) and the system spring constant (taking into account the spring constants of the cantilever and the studied complex). The same range of the retraction velocity (1.5–2.9 μm/s) applied during the measurements, two distinct areas of loading rate variations were obtained i.e., from 2500 to 5000 pN/s and from 3600 to 6600 pN/s for HCV29 and T24 cells, respectively. The corresponding changes of the unbinding forces varied between 25.3 ± 8.2 pN and 27.9 ± 7.5 pN (for reference HCV29 cells) and of 59.7 ± 13.0 pN and 64.1 ± 10.6 pN (cancerous T24 cells). Since the unbinding force variations are comparable with the experimental noise level (7 pN), one can conclude that there is almost no (or exists a weak) dependency between the unbinding force and loading rate. In such case, N-cadherin interactions occurring for both cell types can be compared directly. It would be difficult when strong loading rate dependence would be observed.

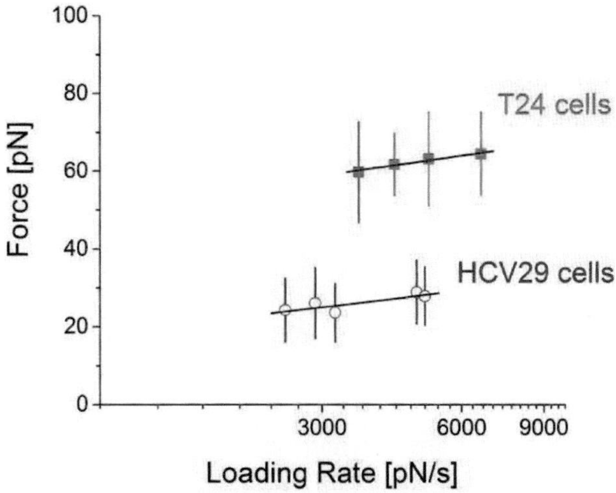

Figure 5.27 Loading rate dependence for Ncadh–GC4 interaction determined for HCV29 and T24 cells in the range of 2500–6600 pN/s. The observed force variations were within the experimental errors, thus, the increasing tendency (line) cannot be relied upon or estimation of interaction kinetics. Reprinted with permission from [54].

5.3.1.3 Force histograms for Ncad–GC4 complex

The force histograms of the interaction force measured in both cell types are presented in Fig. 5.28. The observed peaks were fitted with the Gaussian function in order to determine the value of the most probable unbinding force.

The half width at the half maximum denotes the standard deviation (SD). The rupture of the GC4–Ncadh complex in the reference cells (HCV29) results in the histogram (Fig. 5.28a) with four distinct peaks centered at 26.1 ± 7.1 (SD) pN, 56.9 ± 11.6 (SD) pN, 86.3 ± 6.3 (SD) pN, and 108.3 ± 8.8 (SD) pN, while in the cancerous T24 cells (Fig. 5.28b), one dominant maximum was observed centered at 61.7 ± 14.6 (SD) pN and a small, second peak was visible at 112.2 ± 10.9 (SD) pN. The interaction force of a single Ncadh–GC4 complex was expected to have the same value for both cell types (if the structure of the N-cadherin binding site was unchanged upon cancer transformation). However, the experimental results were clearly different. The unbinding force studied for N-cadherins present in cancer cells (T24) was about 2.5 times higher than in case when cadherins were probed on a surface of the reference HCV29 cells (61.7 ± 14.6 (SD) pN versus 26.1 ± 7.1 (SD) pN). These results indicated an altered structure of the binding site of the cadherins, influencing the binding stability.

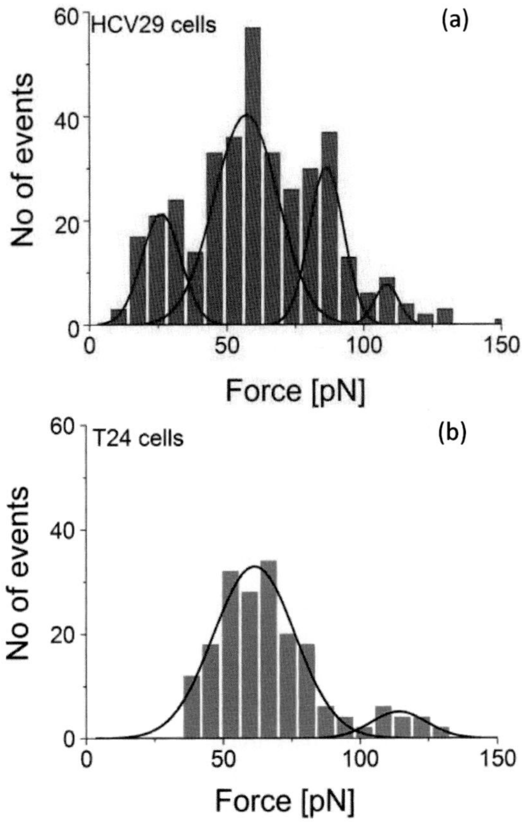

Figure 5.28 Force histogram of the GC4-Ncadh interaction measured for (a) reference HCV29 and (b) cancerous T24 cells. The bin size was 7 pN corresponding to the force detection limit in the experiment (the solid line denotes the fitted Gaussian functions. Reprinted with permission from [54].

5.3.1.4 Multiple unbinding in human bladder cells

The presence of multiple peaks in force histograms can be translated into the relationship of the unbinding force determined for each consecutive peak and the number of bonds simultaneously ruptured (Fig. 5.29).

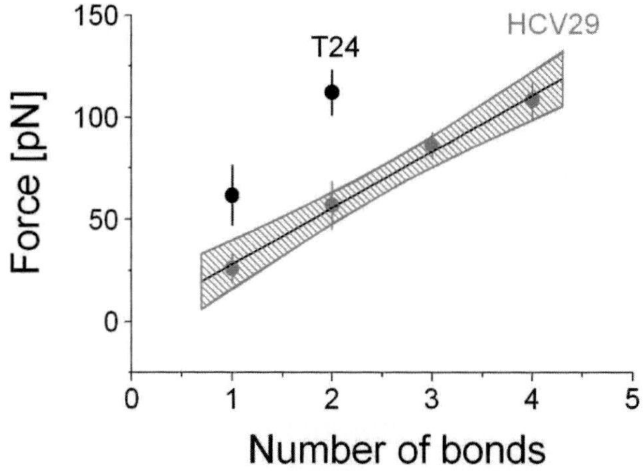

Figure 5.29 Unbinding force as a function of number of simultaneously ruptured bonds. Data points correspond to centers of Gaussians fitted to the corresponding peaks present in force histograms. The 95% confidence bands are marked as grey region. Reprinted with permission from [2].

The linear character is expected when the distance between subsequent maxima is equal, which indicates that only one type of interaction is present. Such relation indicates the specificity of the obtained unbinding events. The slope of the fitted line was 28.1 ± 3.5 pN (the error is the standard deviation of the slope), which corresponds well with the position of the first maximum in the histogram (26.1 ± 7.1 (SD) pN). Therefore, this value was attributed to the unbinding of a single bond. Such analysis was not possible for N-cadherins probed on the surface of T24 cancer cells since only two maxima were observed. Nevertheless, the positions of the first and the second peaks (second doubles first) strongly suggest the discrete character of the observed unbinding, which is characteristic for the specific molecular interactions.

The area under each peak in a force histogram is related to the probability of simultaneous rupture of a given number of single complexes. The results of the rupture of the Ncadh–GC4 complex showed that in case of reference cells the simultaneous unbinding of two single complexes was dominant (probability of ~0.6, Fig. 5.30).

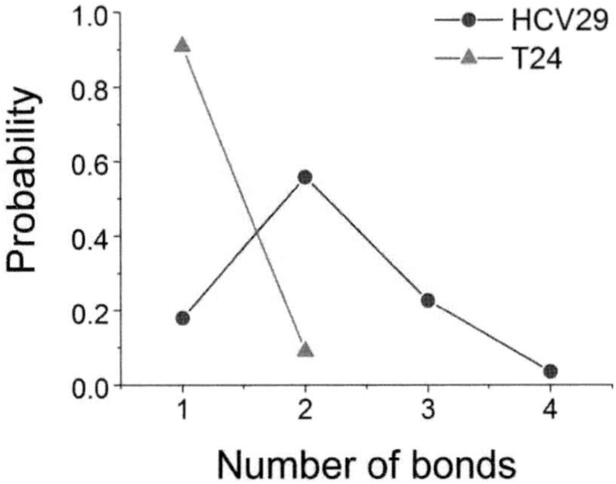

Figure 5.30 Frequency occurrence related to the probability of the unbinding of *n*-bonds for reference, non-malignant HCV29 and cancerous T24 cells. Reprinted with permission from [2].

The probability of the unbinding of 1 or 3 bonds was around 0.2. The simultaneous unbinding of 4 bonds was still probable at the very low level of about 0.02. For cancerous cells, the unbinding of a single pair of molecules was dominant (0.91); however, a small fraction (0.09) of simultaneous rupture of two bonds was also observed.

The positions of the force peaks, plotted as a function of the number of broken bonds, can be compared with two models describing the mechanism of the unbinding "*zipper-like*" and "*parallel-like*" ones (Fig. 5.31).

The general character of the unbinding of GC4–Ncadh complex in both cell types followed the "*parallel-like*" unbinding mechanism in which the applied external force is equally distributed to all single complexes ruptured at the same moment. These results showed also the cooperative character of the N-cadherin unbinding, indicating the similarity to the interaction, recently reported for E-cadherin [60]. However, in HCV29 cells, such mechanism of the unbinding was strongly suggested only after the rupture of 3 or 4 single complexes. For 2 complexes, the "*zipper-like*" model could be applied also with relative high accuracy. In cancerous T24 cells, the simultaneous rupture of two single complexes showed disagreement with "zipper-like" model.

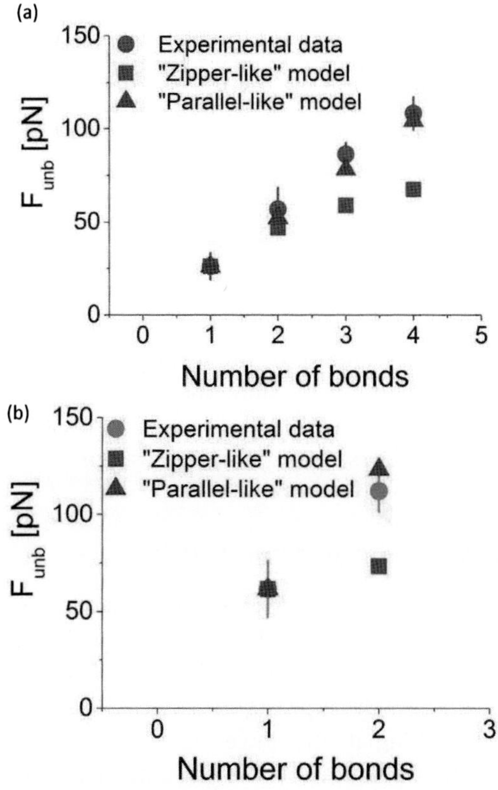

Figure 5.31 The experimental data compared with two theoretical models describing the mechanism of the unbinding: "*zipper-like*" and "*parallel-like*" for Ncadh–GC4 complex in (a) HCV29 and (b) T24 cells. Reprinted with permission from [2].

5.3.1.5 Bell–Evans model parameters

The Bell–Evans model gave a theoretical framework for the understanding how the force affects the dissociation pathway of the molecular complex [7, 9]. The linear fit to the data presented in Fig. 5.27 delivered the slope and intercept, from which the Bell–Evans model parameters were calculated according to Eqs. 5.35 and 5.36 (Table 5.3).

Table 5.3 The Bell model parameters of the Ncadh–GC4 interaction

Cell type	Loading rate (pN/s)	x_b (Å)	k_0 (s^{-1})	t_0 (s)	ΔG (k_BT)
HCV29 (reference)	2500–5000	6.8 ± 0.3	7.8 ± 14.7	0.13 ± 0.25	75.8 ± 1.9
T24 (cancer)	3600–6600	6.3 ± 0.1	0.05 ± 0.01	20.0 ± 2.9	80.8 ± 0.1

Note: Results are presented as a mean ± maximum error (MaxEr). ΔG was calculated according to Eq. (5.23).

The large discrepancy in the unbinding forces, i.e., 26.1 ± 7.1 (SD) pN (HCV29 cells) and 61.7 ± 14.6 (SD) pN (T24 ones), indicates the altered stability of the formed complex, since easier unbinding requires less force. Only one linear region observed in Fig. 5.27 indicates that, within the range of the applied loading rates, only the outermost energy barrier was present in both cell lines.

5.3.1.6 Energy landscape reconstruction

The single-molecule studies enable to characterize the inter-molecular potential of a single ligand–receptor complex. Without the external force, the dissociation of a complex is governed by its activation energy. By applying an external force, the potential is modified by tilting it and lowering the energy barrier. Such action alters the kinetics of the system. The extent of this change depends on the nature of the intermolecular potential of the dissociating complex [9].

The shape of the energy landscapes can be reconstructed basing on parameters derived from the Bell–Evans model. The reconstructed intermolecular potentials of the Ncadh–GC4 complex probed in the HCV29 and T24 cells are presented in Fig. 5.32.

As it was expected for the same type of molecular complex, very similar energy barrier heights of 75.8 ± 1.9 (MaxEr) k_BT and

80.9 ± 0.1 (MaxEr) k_BT and energy barrier positions of 6.8 ± 0.3 (MaxEr) Å and 6.3 ± 0.1 (MaxEr) Å were obtained for HCV29 and T24 cells, respectively. However, despite these similarities, the dissociation rate constants were totally different in both cell lines. The lifetimes of single Ncad–GC4 complexes, calculated as an inverse of the dissociation rate k_0, were 0.13 ± 0.25 (MaxEr) s and 20.8 ± 2.90 (MaxEr) s, correspondingly. The large discrepancy in molecular complex lifetimes cannot be fully explained by differences in the energy barrier properties since, in this case, the energy barrier height differs only by about 5·k_BT and the energy barrier position of about 0.5 Å.

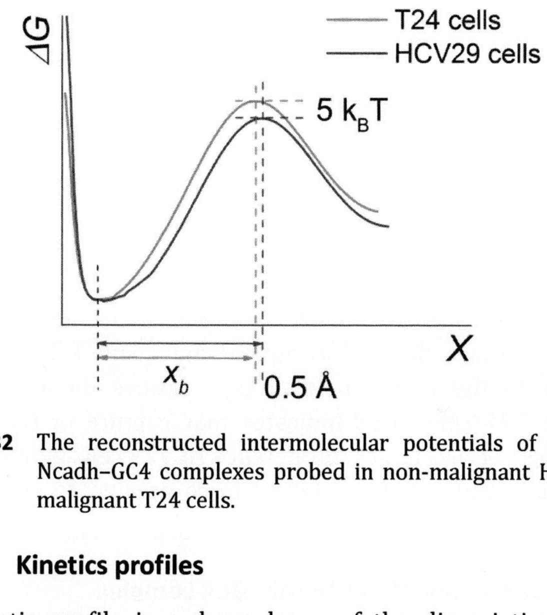

Figure 5.32 The reconstructed intermolecular potentials of the single Ncadh–GC4 complexes probed in non-malignant HCV29 and malignant T24 cells.

5.3.1.7 Kinetics profiles

The kinetic profile is a dependence of the dissociation rate on the external force applied to the bond. Such relation determines how long a bond (or a complex) will last under the changing force. It is analytically described by Eq. (5.9). Based on the determined x_b and k_0 parameters, the kinetic profiles were plotted for all studied complexes. The slope of the kinetic profile is largely determined by the width of the energy barrier. Its larger values result in a steeper slope in the profile.

In order to characterize the differences in the kinetics of Ncadh–GC4 complex in non-malignant HCV29 and malignant

T24 cells, the direct comparison of the dissociation kinetics was performed (Fig. 5.33).

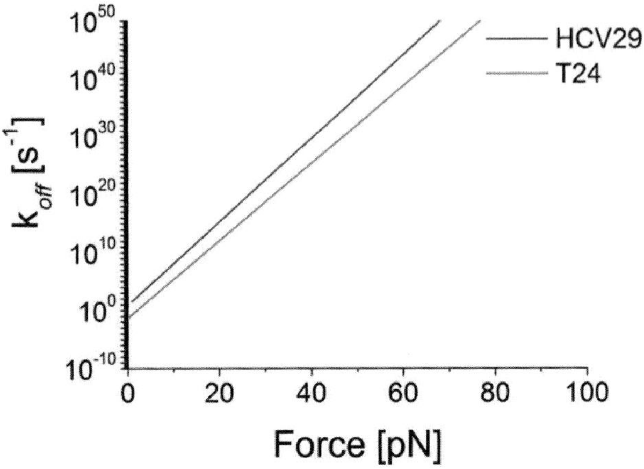

Figure 5.33 The comparison of kinetic profiles of the dissociation of the Ncadh–GC4 complex probed on the surface of non-malignant HCV29 and malignant T24 cells.

The rupture of single Ncadh–GC4 complexes shows that dissociation rate for Ncadh–GC4 complex to be larger when it is probed on a surface of living non-malignant HCV29 cells as compared to the same complex type probes on a surface of malignant T24 cells. This indicates that rupture of the Ncadh–GC4 complex occurs easier in reference HCV29 cells independently on the magnitude of the external force applied to rupture the complex.

5.3.1.8 Specificity of the Ncadh–GC4 complex

The inhibition of molecular interactions with the use of monoclonal antibodies is often used to prove the specific character of the measured unbinding forces. To assure this, cells are incubated with the freely dissolved molecules of the same type as those attached to the AFM probe. In this manner, freely dissolved molecules bind to some of the N-cadherin binding sites, thereby competing with antibody molecules attached to the AFM probe. As a consequence, the number of recorded curves showing the unbinding events decreases (Fig. 5.34a).

Thus, for Ncadh–GC4 complex, after blocking the N-cadherins on the cell surface, the total number of unbinding events, expressed by the unbinding probability value, dropped from 0.036 to 0.005 and from 0.088 to 0.021 for HCV29 and T24 cells, respectively [54]. The unbinding probability correlates with the number of binding sites present on a surface and participating in the binding process. The lower unbinding probability value indicates a smaller number of binding sites available for the antibody recognizing N-cadherins on a cells surface. The unbinding probability for the studied cells was 0.036 (HCV29) and 0.088 (T24), which agrees qualitatively, with the Western blot results showing the overexpression of N-cadherin in cancerous T24 cells (Fig. 5.34b).

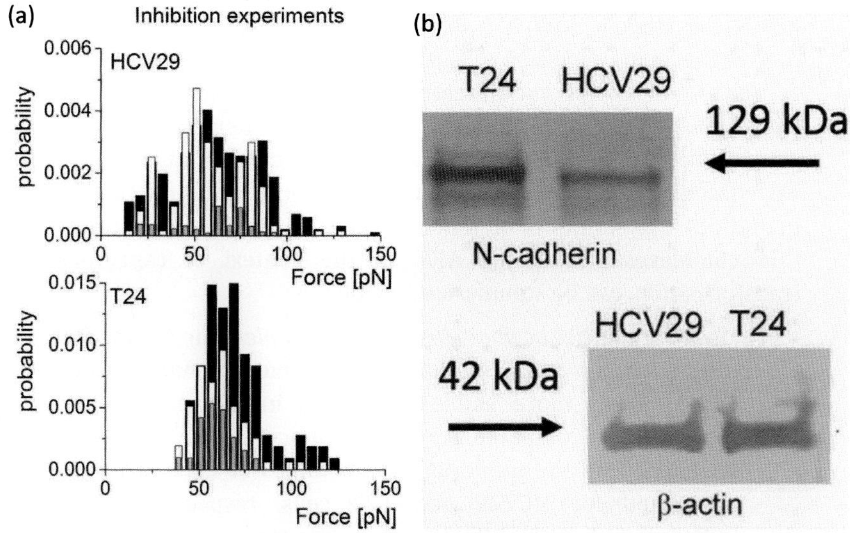

Figure 5.34 (a) The inhibition of the Ncadh–GC4 interaction. The number of unbinding events decreases as concentration of blocking antibody increases. Black columns—unbinding measurements without inhibition; white columns—events after the inhibition at the GC4 antibody concentration of 0.1 mg/ml; gray columns—events after the inhibition at 0.4 mg/ml GC4 antibody concentration. All histograms were normalized by the total number of force curves recorded. (b) The Western blot analysis showing the overexpression of N-cadherin in cancerous T24 cells. Reprinted with permission from [54].

5.3.1.9 Summary for Ncadh–CG4 complex

The section 5.3 describes only one type of the interaction occurring between N-cadherin and monoclonal antibody GC4 possessing binding site between two first ectodomains of N-cadherin. The focus on only one molecular complex was settled to demonstrate the great capability of the AFM-based unbinding force measurements in the characterization of unbinding properties at a single molecular level. The level of the expression of molecules, determined by AFM, is limited only to molecules present in a plasma membrane and it can be directly related to the classical way of the molecule expression determination through the unbinding probability value: a parameter describing the number of molecules on cell surface. The obtained results are in good agreement with indirect (and rather qualitative) results which have been already reported pointing out an increased expression of N-cadherin in cancerous cells [61].

The analysis of the interaction between a given pair of single molecules performed by AFM allows determination of sets of parameters that more precisely and in a quantitative way describe the expression of molecules on a cell surface. The quantitative description of the molecule expression for Ncadh–GC4 complexes studied by AFM in the context of cancerous transformation, can be summarized as follows:

(1) *Static properties of single Ncadh–GC4 complex.* The interaction occurring between N-cadherin and its monoclonal antibody (GC4) forms more stable complexes in cancerous (T24) cells when compared to non-malignant (HCV29) ones: visualized by the unbinding force values of 26 and 61 pN measured for HCV29 and T24 cells, respectively. This difference is attributed to the changes of the antibody binding site, which may reveal the alterations in both the primary N-cadherin structure and/or different glycan pattern in the neighborhood of the binding site.

(2) *Dynamic properties of single molecular Ncadh–GC4 complex.* The Ncadh–GC4 complex dissociates differently depending on cell type studied. The reconstructed energy potential for cancerous cells has almost two times larger barrier than that obtained for the energy landscape of reference cells. The calculated height of the corresponding barriers (the difference was of $5k_BT$, defined by the dissociation rate

constants) is similar. This indicates faintly slower way of the Ncadh–GC4 complex dissociation in cancerous cells. The lifetime of a single GC4–Ncadh complex, determined from the dissociation rate constant, has similar values, which means that, independently of the minimal difference in the energy heights, passing over the energy barrier is only slightly burden by the oncogenic transformation. The calculated kinetic profile indicates that the faster dissociation of Ncadh–GC4 complexes in non-malignant HCV29 cells as compared to cancerous T24 ones.

(3) *Dynamic properties of adhesion clusters.* The mechanism of unbinding of clusters of Ncadh–GC4 complexes in both cell types followed the "*parallel-like*" unbinding mechanism. These results showed the cooperative character of the N-cadherin unbinding indicating the similarity to the interaction for E-cadherin.

So far there is no data reporting the different structure of N-cadherins in cancer and normal cells. Therefore, one may assume that the primary structure of N-cadherin in both non-malignant (HCV29) and malignant (T24) cells is the same. Consequently, the difference of the unbinding forces can be attributed to structural changes of oligosaccharide ligands caused by cancer progression, especially since the different glycosylation pattern was already observed for these cell lines [62]. As T24 cancer cells have ability to generate more complex glycans attached to cadherins than HCV29 cells, one should expect smaller unbinding force in T24 due to the presence of large amount of attached glycans generating a steric-originated repulsion, preventing cadherins from achieving distances necessary for the effective interaction. However, for the T24 cells, a larger unbinding force was observed in the experiment than for the HCV29 cells. This result might indicate the smaller amount of glycans present in the neighborhood of the antibody (GC4) binding site and thereby less prominent steric repulsion.

5.4 Living Cell as a Probe

When a single cell is used as a probe (referred here as a *cell probe*), the recorded force curves are far more complex as compared to those observed in single-molecule experiments. Examples of curves recorded using a single WM115 melanoma cell interacting with polystyrene-coated glass surface are presented in Fig. 5.35a–f.

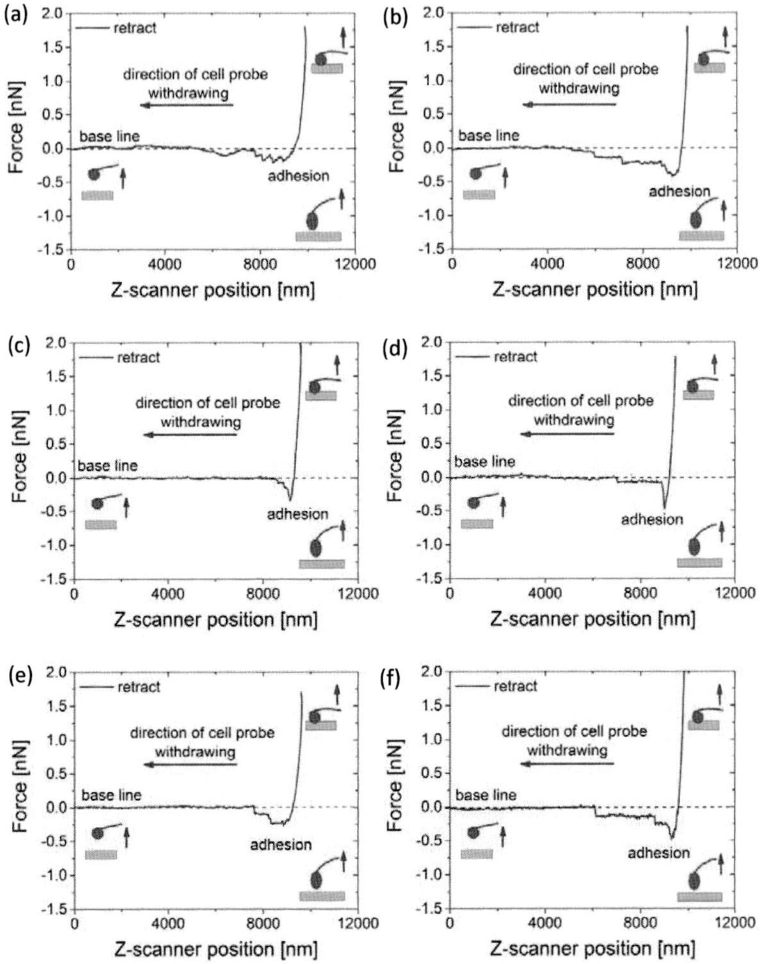

Figure 5.35a–f Examples of force curves recorded using a single WM115 melanoma cell as a probe representing various shapes thereby indicating distinct interaction force acting between a cell and polystyrene-coated glass surface. Unpublished data, courtesy of Szymon Prauzner-Bechcicki, IFJ PAN.

Force curves were recorded over a scan area of 0.5 μm × 0.5 μm with the retraction speed of 8 μm/s and a grid of

8 × 8 pixels. The distance between two subsequent probing locations was 62.5 nm, much smaller than the size of a probing single cell. One can expect that majority of the curves would have similar shape characteristic for the interaction between the same, individual, single WM115 cell and polystyrene coated glass surface. On the contrary, each of the recorded curves has different shape of the adhesion-related part, indicating a huge variety of the possible interaction forces acting within a contact area of a cell and a polystyrene surface. The adhesion, observed during the cell unbinding, contains contributions from various distinct processes like sequential and parallel ruptures of molecular bonds, viscous and elastic deformations of the cell body and membrane or non-specific interactions forces. To quantify the adhesion several quantities can be derived from the analysis of force curve recorded for cell-surface (or cell-cell) interactions (Fig. 5.36).

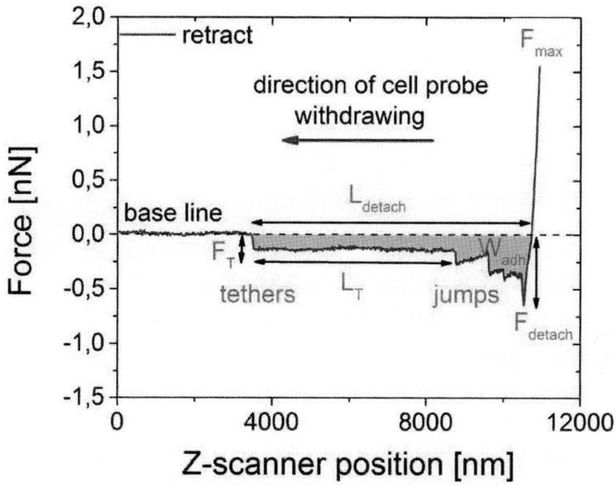

Figure 5.36 The way of quantitative analysis of the retraction part of the force curve recorded using a single cell as a probe. The overall adhesion can be quantify by a work of adhesion (W_{adh}), a distance and a force at which detachment occurs (L_{detach}, F_{detach}), respectively. The tethers and single jumps can be described by their length and force at which they detach from the surface (for tethers: L_T, F_T).

When the cell is pressed against the surface, the force increases until it reaches a pre-set force level (F_{max}). Afterwards, the cell is withdrawn from the surface. Any bonds that have been formed during contact break until the cell has completely separated from the surface (*base line* in Fig. 5.36). The overall adhesion magnitude can be quantified by a work of adhesion (shaded area in Fig. 5.36), the detachment force (F_{detach}), defined as a maximum force needed to separate cell and surface (or cell and cell), and a distance at which the detachment occurs (L_{detach}). The force plateau (Fig. 5.36) is usually attributed to tethers extruded from cell membrane. They can be quantified by the force needed to rupture a single tether (F_T) and tether's length (L_T). Each single jump (present as step-like unbinding event) can be also quantified by rupture force and distance values.

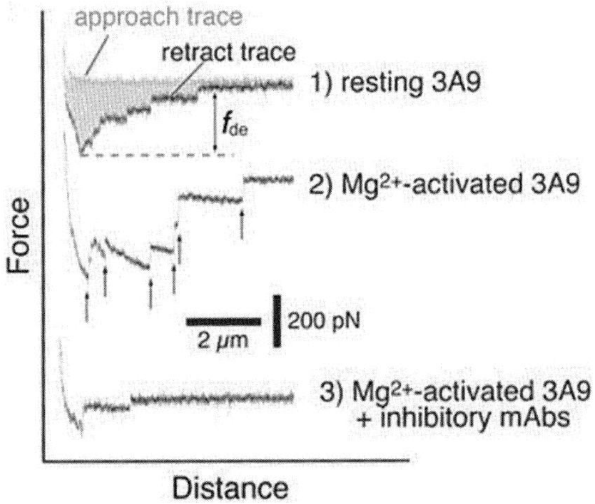

Figure 5.37 The AFM measurements of the interaction between 3A9 cells (used as the AFM cell probe) and ICAM-1 molecules immobilized on a surface of culture dish: (1) resting cell, (2) Mg^{2+}-treated, and (3) monoclonal antibody inhibited interactions. The shaded area in the case (1) corresponds to the work of adhesion needed to detach the 3A9 cell from the ICAM-1 coated surface. The detachment force f_{de} is supported by the adhesive bonds formed between the cell and the substrate. The arrows in case (2) point to positions in the force curve where the formed complexes ruptured. The addition of monoclonal antibody inhibited the studied interaction. Reprinted with permission from [45].

In one of the first papers, cell probes have been used to study the interaction between T-cells and ICAM-1 functionalized surface [45]. The ICAM-1 molecule is a cell surface glycoprotein that, together with the leucocyte function-associated antigen-1 (LFA-1), modulates lymphocyte function. The interaction between ICAM-1 and LFA-1 requires the presence of divalent ions, such as Mg^{2+}, facilitating the binding. An exemplary force curves recorded for the interaction between ICAM-1 and LFA-1 molecules detected in living 3A9 cells (murine CD4 T-cell hybridoma) is presented in Fig. 5.37.

All curves of the 3A9 cell interaction with ICAM-1 coated surface were acquired with the same cell, under the same experimental conditions (300 pN compression force, 5s contact, and 2 μm/s retraction speed) in the following groups: (1) resting, (2) Mg^{2+} activated, and (3) Mg^{2+} activated and inhibited by monoclonal antibody cells.

The interaction between 3A9 cell and ICAM-1 molecules was characterized by detachment work and force used here as measures of cell adhesion (Fig. 5.38).

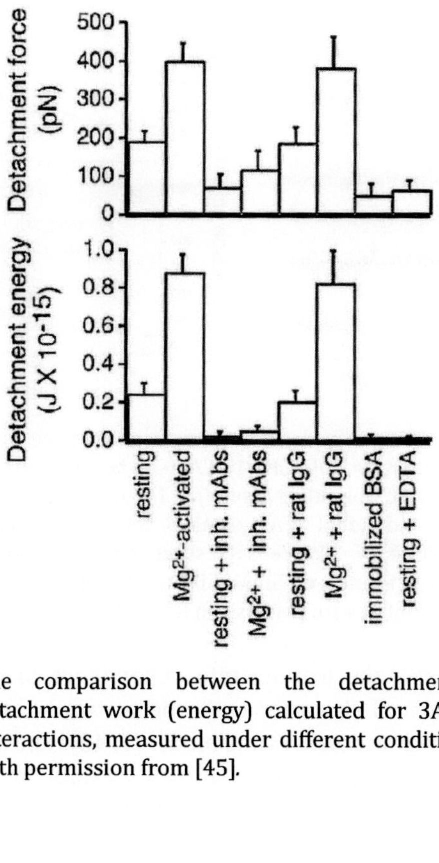

Figure 5.38 The comparison between the detachment force and detachment work (energy) calculated for 3A9 cell–ICAM-1 interactions, measured under different conditions. Reprinted with permission from [45].

Work of adhesion is directly correlated to energy that cell dissipates during the detachment. For the A39 cell–ICAM-1 molecule interaction, independently of the derived quantity, i.e., detachment force or work (energy), the qualitative relation is similar.

From the single molecule experiments, it is known that the unbinding force is strongly dependent on the loading rate. In case of measurements carried out on living cells, finding the curves showing characteristics events that are similar to single molecule unbinding is not an easy task. This can be realized by minimizing the contact between a cell and a surface, i.e., by reducing both contact time and compression force. For the case of 3A9 cells and ICAM-1 coated surface, such simply reduction manifests in the appearance of force curves showing unbinding events characteristic for single molecule interaction (Fig. 5.39a).

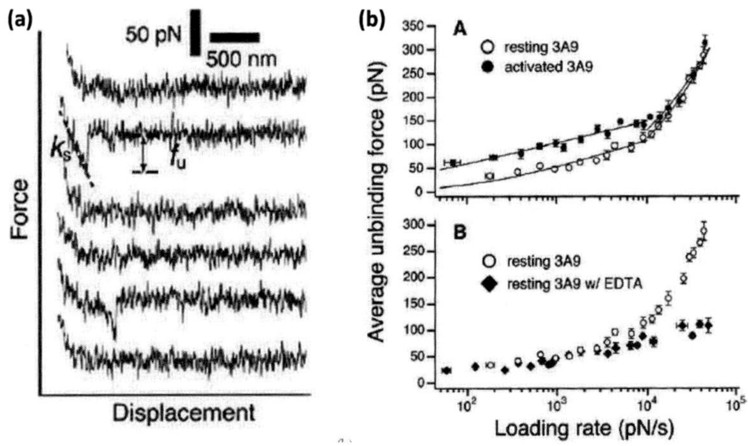

Figure 5.39 (a) Force-displacement curves showing single molecule interaction of LFA-1/ICAM-1 bond characterized by a specific unbinding force (f_u). (b) Loading rate dependence of the individual LFA-1/ICAM-1 force measured directly on a surface of 3A9 cells (resting, Mg^{2+} activated cells, and in the presence of 5mM EDTA) showing two linear regimes. Reprinted with permission from [45].

In such case, the loading rate dependence of the unbinding force shows typical character with one or two linear regions that can be fitted with the Bell–Evans model [9, 45]. The measurements demonstrated that LFA-1/ICAM-1 bonds exhibit different kinetic behavior at slow (below 10^4 pN/s) and fast (above this rate) loading regimes. This indicates that the dissociation of the complex involves overcoming a steep, inner barrier and a wide, outer activation one.

For all cases, where it is not possible to observe single molecule interactions, adhesion forces plotted as a function of retraction velocity (which is indirectly related to the loading rate value) can help to evaluate kinetics of single cell unbinding. Recently, such approach was used to study the adhesion force between an endothelial cell monolayer and human bladder cancer cells (Fig. 5.40) [63].

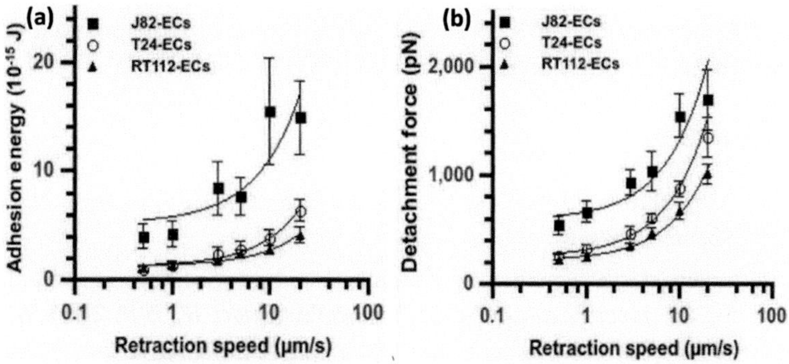

Figure 5.40 The adhesion energy and detachment force plotted as a function of retraction speed determined for the interaction between human bladder cancer (J82, T24 and RT112 cell lines) and HUVEC endothelial cells (ECs). Analogously as for lymphocytes (Fig. 5.38), both the adhesion energy and the detachment force represent similar dependency. Reprinted with permission from [63].

Three bladder cell lines were compared: RT112, T24, and J82. These cell lines represent distinct stages of cancer progression from well to poorly differentiated phenotypes. RT112 cancer cells are moderately differentiated cancer. T24 and J82 cancer cells are poorly differentiated cancers. The interaction of cancer cells to endothelial HUVEC cells (*Human Umbilical Vascular Endothelial Cells*) was characterized by both the adhesion energy and detachment force, plotted as a function of retraction speed. As shown in Fig. 5.40, these two parameters increase with retraction speed. Regarding the adhesion energy, the most invasive J82 cells present larger values as compared to the T24 or RT112 cells.

This difference is confirmed by the detachment force values which were larger for J82 cells as compared to T24 and RT112 ones. As a further step, by using ICAM-1 coated substrates and a monoclonal antibody specific for ICAM-1, it has been demonstrated that ICAM-1 serves as a key receptor on endothelial cells whose interaction with ligands expressed by cancer bladder cells is correlated with the rupture forces obtained with the most invasive cancer cells (T24, J82). For the less invasive cancer cells (RT112), endothelial ICAM-1 does not seem to play any role in the adhesion process. A detailed analysis suggests that ICAM-1 interacts preferentially with one ligand on T24 cancer cells and with two ligands on J82 cancer cells. Potential receptors, involved in the interaction, are CD43 and MUC1 molecules expressed in the studied human bladder cancer cells [63].

References

1. Bongrand, P. (1999). Ligand–receptor interactions. *Rep. Prog. Phys.*, **62**, pp. 921–968.
2. Lekka, M. (2007). The use of atomic force microscopy as a technique for the identification of cancerous cells. *Report IFJ PAN*, **2001/AP**, pp. 1–147.
3. Orsello, C. E., Lauffenburger, D. A., and Hammer, D. A. (2001). Molecular properties in cell adhesion: A physical and engineering perspective. *Trends Biotechnol.*, **19**, pp. 310–316.
4. Dembo, M., Torney, D. C., Saxman, K., and Hammer, D. (1988). The reaction-limited kinetics of membrane-to-surface adhesion and detachment. *Proc. R. Soc. London Biol. Sci.*, **234**, pp. 55–83.
5. Kramers, H. A. (1940). Brownian motion in a field of force and the diffusion model of chemical reactions. *Physica*, **VII**, pp. 284–304.
6. Hänggi, P., Talkner, P., and Borkovec, M. (1990). Reaction-rate theory: Fifty years after Kramers. *Rev. Mod. Phys.*, **62** pp. 251–341.
7. Bell, G. I. (1978). Models for the specific adhesion of cells to cells. *Science*, **200**, pp. 618–627.
8. Evans, E. (1999). Looking inside molecular bonds at biological interfaces with dynamic force spectroscopy. *Biophys. Chem.*, **82**, pp. 83–97.
9. Evans, E., and Ritchie, K. (1997). Dynamic strength of molecular adhesion bonds. *Biophys. J.*, **72**, pp. 1541–1555.

10. Tees, D. F., Chang, K. C., Rodgers, S. D., and Hammer, D. A. (2002). Simulation of cell adhesion to bioreactive surfaces in shear: The effect of cell size. *Ind. Eng. Chem. Res.*, **41**, pp. 486–493.

11. Fritz, J., Katopodis, A. G., Kolbinger, F., and Anselmetti, D. (1998). Force-mediated kinetics of single P-selectin/ligand complexes observed by atomic force microscopy. *Proc. Natl. Acad. Sci. U. S. A.*, **95**, pp. 12283–12288.

12. Nishizaka, T., Seo, R., Tadakuma, H., Kinosita, K., and Ishiwata, S. (2000). Characterization of single actomyosin rigor bonds: Load dependence of lifetime and mechanical properties. *Biophys. J.*, **79**, pp. 962–974.

13. Kienberger, F., Kada, G., Mueller, H., and Hinterdorfer, P. (2005). Single molecule studies of antibody–antigen interaction strength versus intra-molecular antigen stability. *J. Mol. Biol.*, **347**, pp. 597–606.

14. Neuert, G., Albrecht, C., Pamir, E., and Gaub, H. E. (2006). Dynamic force spectroscopy of the digoxigenin–antibody complex. *FEBS Lett.*, **580**, pp. 505–509.

15. Izrailev, S., Stepaniants, S., Balsera, M., Oono, Y., and Schulten, K. (1997). Molecular dynamics study of unbinding of the avidin-biotin complex. *Biophys. J.*, **72**, pp. 1568–1581.

16. Evans, E. (2001). Probing the relation between force-lifetime and chemistry in single molecular bonds. *Annu. Rev. Biophys. Biomol. Struct.*, **30**, pp. 105–128.

17. Li, F., Redick, S. D., Erickson, H. P., and Moy, V. T. (2003). Force measurements of the $\alpha_5\beta_2$ integrin-fibronectin interaction. *Biophys. J.*, **84**, pp. 1252–1262.

18. Zhang, X., Bogorin, D. F., and Moy, V. T. (2004). Molecular basis of the dynamic strength of the sialyl Lewis X–selectin interaction. *ChemPhysChem*, **5**, pp. 175–182.

19. Lebed, K., Kulik, A. J., Forro, F., and Lekka, M. (2006). Lectin–carbohydrate affinity measured using quartz crystal microbalance. *J. Coll. Interface Sci.*, **299**, pp. 41–48.

20. Homola, J. (2003). Present and future of surface plasmon resonance biosensors. *Anal. Bioanal. Chem.*, **377**, pp. 528–539.

21. Jelesarov, I., and Bosshard, R. H. (1999). Isothermal titration calorimetry and differential scanning calorimetry as complementary tools to investigate the energetics of biomolecular recognition. *J. Mol. Recognit.*, **12**, pp. 3–18.

22. Strunz, T., Oroszlan, K., Schaefer, R., and Gunterodt, H. J. (1999). Dynamic force spectroscopy of single DNA molecules. *Proc. Natl. Acad. Sci. U. S. A.*, **96**, pp. 11277–11282.
23. Cocco, S., Monasson, R., and Marko, J. F. (2001). Force and kinetics barriers to initiation of DNA unzipping. *Phys. Rev. E*, **65**, p. 0419071.
24. Williams, P. M. (2003). Analytical descriptions of dynamic force spectroscopy: Behavior of multiple connections. *Anal. Chim. Acta*, **479**, pp. 107–115.
25. Seifert, U. (2000). Rupture of multiple parallel bonds under dynamic loading, *Phys. Rev. Lett.*, **84**, pp. 2750–2754.
26. Sulchek, T. A., Friddle, R. W., Langry, K., Lau, E. Y., Albrecht, H., Ratto, T. V., DeNardo, S. J., Colvin, M. E., and Noy, A. (2005). Dynamic force spectroscopy of parallel individual mucin1–antibody bonds. *Proc. Natl. Acad. Sci. U. S. A.*, **102**, pp. 16638–16643.
27. Bizzarri, A. R., and Cannistraro, S. (2012). *Dynamic Force Spectroscopy and Biomolecular Recognition*, 1st ed. (CRC Press Taylor & Francis Group, Boca Raton, USA).
28. Hummer, G., and Szabo, A. (2003). Kinetics from non-equilibrium single molecule pulling experiments. *Biophys. J.*, **85**, pp. 5–15.
29. Dutko, O. K., Hummer, G., and Szabo, A. (2006). Intrinsic rates and activation free energies from single-molecule pulling experiments. *Phys. Rev. Lett.*, **96**(1–4), p. 108101.
30. Friddle, R. W., Noy, A., and De Yoreo, J. J. (2012). Interpreting the widespread nonlinear force spectra of intermolecular bonds. *Proc. Natl. Acad. Sci. U. S. A.*, **109**, pp. 13573–13578.
31. Senkara-Barwijuk, E. Lebed, K., Kobiela, T., and Lekka, M. (2012). Reaction pathway and free energy profile determined for specific recognition of oligosaccharide moiety of carboxypeptidase Y. *Biosens. Bioelectron.*, **36**, pp. 103–109.
32. Merkel, R., Nassoy, P., Leung, A., Ritchie, K., and Evans, E. (1999). Energy landscapes of receptor-ligand bonds explored with dynamic force spectroscopy. *Nature*, **397**, pp. 0–53.
33. Pierres, A., Benoliel, A. M., Bongrand, P., and van der Merwe, P. A. (1996). Determination of the lifetime and force dependence of interactions of single bonds between surface-attached CD2 and CD48 adhesion molecules. *Proc. Natl. Acad. Sci.*, **93**, pp. 15114–15118.
34. Smith, S. B., Finzi, L., and Bustamante, C. (1992). Direct mechanical measurements of the elasticity of single DNA molecules by using magnetic beads. *Science*, **258**, pp. 1122–1126.

35. Kellermayer, M. S. (2005). Visualizing and manipulating individual protein molecules. *Physiol. Meas.*, **26**, pp. R119–R153.

36. Allen, S., Davies, J., Davies, M. C., Dawkes, A. C., Roberts, C. J., Tendler, A. J. B., and Williams, P. M. (1999). The influence of epitope availability on atomic-force microscope studies of antigen-antibody interactions. *Biochem. J.*, **341**, pp. 173–178.

37. Hinterdorfer, P., and Dufrene, Y. F. (2006). Detection and localization of single molecular recognition events using atomic force microscopy. *Nat. Methods*, **3**, pp. 347–355.

38. Helenius, J., Heisenberg, C. P., Gaub, H. E., and Muller, D. J. (2008). Single-cell force spectroscopy. *J. Cell. Sci.*, **121**, pp. 1785–1791.

39. Hinterdorfer, P., Baumgartner, W., Gruber, H. J., Schilcher, K., and Schindler, H. (1996). Detection and localization of individual antibody-antigen recognition events by atomic force microscopy. *Proc. Natl. Acad. Sci. U. S. A.*, **93**, pp. 3477–3481.

40. Verbelen, C., Gruber, H. J., and Dufrene, Y. F. (2007). The NTA-His6 bond is strong enough for AFM single-molecular recognition studies. *J. Mol. Recognit.*, **20**, pp. 490–494.

41. Dupres, V., Menozzi, F. D., Locht, C., Clare, B. H., Abbott, N. L., Cuenot, S., Bompard, C., Raze, D., and Dufrene, Y. F. (2005). Nanoscale mapping and functional analysis of individual adhesins on living bacteria. *Nat. Methods*, **2**, pp. 515–520.

42. Gabai, R., Segev, L., and Joselevich, E. (2005). Single polymer chains as specific transducers of molecular recognition in scanning probe microscopy. *J. Am. Chem. Soc.*, **127**, pp. 11390–1398.

43. Israelaschvili, J. N. (1992). *Intermolecular and Surface Forces*, 2nd ed. (Academic Press Inc., San Diego, USA).

44. Zlatanova, J., and vanHolde, K. (2006). Single-molecule biology: What is it and how does it work?. *Mol. Cell*, **24**, pp. 317–329.

45. Zhang, X., Wojcikiewicz, E., and Moy, V. T. (2002). Force Spectroscopy of the leukocyte function-associated antigen-1/intercellular adhesion molecule-1 interaction. *Biophys. J.*, **83**, pp. 2270–2279.

46. Lebed, K., Kulik, A. J., Forró, L., and Lekka, M. (2007). Study the kinetics of lectin carbohydrate interaction using atomic force microscopy and quartz crystal microbalance methods. *Acta Phys. Polon. A*, **111**, pp. 273–287.

47. Lebed, K., Dąbrowska, A., Sharma, G., Lekki, J., Stachura, Z., and Lekka, M. (2006). Force spectroscopy of mannose-binding to patterned

concanavalin A and lentil lectin. *Proc. VIII Linz Winter Workshop*, pp. 84–86.

48. Goffin, J. M., Pittet, P., Csucs, G., Lussi, J. W., Meister, J. J., and Hinz, B. (2006). Focal adhesion size controls tension-dependent recruitment of α-smooth muscle actin to stress fibers. *J. Cell. Biol.*, **172**, pp. 259–268.
49. Lebed, K., Kulik, A. J., Forro, L., and Lekka, M. (2004). Lectin–carbohydrate affinity measured using quartz crystal microbalance. *J. Coll. Interface Sci.*, **299**, pp. 41–48.
50. Laidler, P., Dulińska, J., Lekka, M., and Lekki, J. (2005). Expression of prostate specific membrane antigen in androgen–independent prostate cancer cell line PC–3. *Arch. Biochem. Biophys.*, **435**, pp. 1–14.
51. Lekka, M., Laidler, P., Dulińska, J., Łabędź, M., and Pyka, G. (2004). Probing molecular interaction between concanavalin A and mannose ligands by means of SFM. *Eur. Biophys. J.*, **33**, pp. 664–650.
52. Chtcheglova, L. A., and Dietler, G. (2003). Force spectroscopy of polyclonal and monoclonal anti-bovine serum albumin antibodies–BSA complexes. *Acta Phys. Pol.*, **104**, pp. 321–326.
53. Lekka M., Laidler, P., Łabędź, M., Kulik, A. J., Lekki, J. Zając, W., and Stachura, Z. (2006). Specific detection of glycans on a plasma membrane of living cells using atomic force microscopy. *Chem. Biol.*, **13**, pp. 505–512.
54. Lekka, M., Gil, D., Dabros, W., Jaczewska, J., Kulik, A. J., Lekki, J., Stachura, Z., Stachura, J., and Laidler, P. (2011). Characterization of N-cadherin unbinding properties in non-malignant (HCV29) and malignant (T24) bladder cells. *J. Mol. Recognit.*, **24**, pp. 833–842.
55. Fouquet, B., Zimbelmann, R., and Franke, W. W. (1992). Identification of plakoglobin in oocytes and early embryos of Xenopus laevis: Maternal expression of a gene encoding a junctional plaque protein. *Differentiation*, **51**, pp. 87–194.
56. Ramburan, A., and Govender, D. (2002). Cadherins and catenins in pathology. *Curr. Diagn. Pathol.*, **8**, pp. 305–317.
57. Nollet, F., Berx, G., and vanRoy, F. (1999). The role of the E-cadherin/catenin adhesion complex in the development and progression of cancer. *Mol. Cell Biol. Res. Commun.*, **2**, pp. 77–85.
58. Reiss, K., Maretzky, T., Ludwig, A., Tousseyn, T., deStrooper, B., Hartmann, D., and Saftig, P. (2005). ADAM10 cleavage of N-cadherin and regulation of cell-cell adhesion and β-catenin nuclear signaling. *EMBO J.*, **24**, pp. 742–752.

59. Harrison, O. J., Corps, E. M., Berge, T., and Kilshaw, P. J. (2005). The mechanism of cell adhesion by classical cadherins: The role of domain 1. *J. Cell. Sci.,* **118,** pp. 711–721.

60. Tsukasaki, Y., Kitamura, K., Shimizu, K., Iwane, A. H., Takai, Y., and Yanagida, T. (2007). Role of multiple bonds between the single cell adhesion molecules, nectin and cadherin, revealed by high sensitive force measurements. *J. Mol. Biol.,* **367,** pp. 996–1006.

61. Giroldi, L. A., Bringuier, P. P., Shimazui, T., Jansen, K., and Schalken, J. A. (1999). Changes in cadherin–catenin complexes in the progression of human bladder carcinoma. *Int. J. Cancer,* **82,** pp. 70–76.

62. Przybyło, M., Hoja-Łukowicz, D., Lityńska, A., and Laidler, P. (2002). Different glycosylation of cadherins from human bladder non-malignant and cancer cell lines. *Can. Cell Int.,* **2,** pp. 6–10.

63. Laurent, V. M., Duperray, A., Sundar Rajan, V., and Verdier, C. (2014). Atomic force microscopy reveals a role for endothelial cell ICAM-1 expression in bladder cancer cell adherence. *PLoS ONE,* **9,** p. e98034.

Chapter 6

Conclusions

Atomic force microscopy (AFM) is undoubtedly one of the most widely used techniques applied in studies focused on the characterization of physical properties of individual cells in conditions close to natural ones. Experimental data collection, advanced processing, and interpretation place AFM within the group of well-established techniques, extensively used in the laboratory research. Its strength stems both from the high-resolution imaging and from the ability to quantitatively characterize biophysical properties of single living cells. The ongoing technological development brought so far research instruments that allow carrying out highly complex experiments, where AFM delivers unique information on cellular or molecular processes, not always accessible in other techniques.

In cancer biology, cells' capability to deform and to adhere is strongly linked with the stage of a cell. Research data, accumulated so far, reveal large alterations in cellular structures within a cytoskeleton organization and cellular surface and mechanical properties, during cancer progression [1–3]. Therefore, the quantification of cellular capability to adhere and to deform at a single cell level seems to help and to advance knowledge in various aspects, like cancer cell interactions with extracellular matrix following the mechanism of their migration to distant places in the body and a formation of tumor metastatic sites.

Cellular Analysis by Atomic Force Microscopy
Malgorzata Lekka

ISBN 978-981-4669-67-2 (Hardcover), 978-1-315-36480-3 (eBook)
www.panstanford.com

The determination of the unbinding forces for various molecular complexes enables studies of the molecular complex formation and dissociation at a single molecule level. The properties of individual complexes are diversified due to their participation in dynamic processes and, thus, important information obtained from bulk-related methods may not fully explain phenomena occurring at single cell level. Quantitative and high-resolution data obtained from AFM can help to understand how single molecules unbind and how this is related to adhesive properties of single cells. Alterations in the expression and patterns of cell surface receptors, observed in many cancers, demand answer how such changes are related to the varying number of receptors and to spatial distribution or properties of single complexes. This raises the question whether the unbinding characteristics can be used to study the binding site identity (if the unbinding forces show similar value) or dissimilarity (if large differences are observed) at given experimental conditions and whether such information can be used as a maker of cancer progression. The dynamic force spectra (i.e., dependencies of the unbinding force and loading rate) bring more detailed information about how the molecular complexes rupture. Various studies show that genetically modified complexes reveal different kinetics, which gives hope for deeper understanding the adhesion of cancerous cells [4–6].

The determination of elastic properties of living cells or even tissues by AFM seems to be better developed as compared to the unbinding studies. Various data prove that cancer cells are more deformable as compared to normal (or even benign) cells [7, 8]. Most of the studies were carried out on isolated, frequently immortalized cells where an experimenter a priori knows the type of the studied cells, which easily allows to attribute given cellular deformability to the cell type. However, even good knowledge of single cell mechanical properties not always explains satisfactorily the properties of real samples, where cancerous cells interact with neighboring cells of the same or other types. Modeling cells environment by a mixture of two different cell types cultured using stiff or soft 2D or 3D environment introduces more complexity in the experiment and opens new pathways of understanding cells' mechanical properties and their significance in cancer formation [9, 10].

Due to experimental difficulties, tissue measurements by AFM are still rare. Research data show very complex mechanical characteristics that are difficult to be explained and interpreted [11, 12]. Yet another factor is the influence of the extracellular matrix, which is changing its composition and structure depending on the tissue/cell type and stage of cancer progression. Tissue studies will be probably one of the major research directions in the near future.

Finally, I would like to express my opinion that despite large progress in the AFM technique in the last two decades, still substantial scientific and technological development needs to be applied in the detection and characterization of cancer-related changes.

References

1. Paszek, M. J., Zahir, N., Johnson, K. R., Lakins, J. N., Rozenberg, G. I., Reinhart-King, C. A., Margulies, S. S., Dembo, M., Boettiger, D., Hammer, D. A., and Weaver, V. M. (2005). Tensional homeostasis and the malignant phenotype. *Cancer Cell*, **8**, pp. 241–254.
2. Wirtz, D., Konstantopoulos, K., and Searson, P. C. (2011). The physics of cancer: The role of physical interactions and mechanical forces in metastasis. *Nat. Rev. Cancer*, **11**, pp. 512–522.
3. Tung, J. C., Barnes, J. M., Desai, S. R., Sistrunk, C., Conklin, M. W., Schedin, P., Eliceiri, K. W., Keely, P. J., Seewaldt, V. L., and Weaver, V. M. (2015). Tumor mechanics and metabolic dysfunction. *Free Radic. Biol. Med.*, **79**, pp. 269–280.
4. Kienberger, F., Kada, G., Mueller, H., and Hinterdorfer, P. (2005). Single molecule studies of antibody–antigen interaction strength versus intra-molecular antigen stability. *J. Mol. Biol.*, **347**, pp. 597–606.
5. Zhang, X., Bogorin, D. F., and Moy, V. T. (2004). Molecular basis of the dynamic strength of the sialyl Lewis X–selectin interaction. *ChemPhysChem*, **5**, pp. 175–182.
6. Laurent, V. M., Duperray, A., Sundar Rajan, V., and Verdier, C. (2014). Atomic force microscopy reveals a role for endothelial cell ICAM-1 expression in bladder cancer cell adherence. *PLoS ONE*, **9**, e98034.
7. Lekka, M., Pogoda, K., Gostek, J., Klymenko, O., Prauzner-Bechcicki, Sz., Wiltowska-Zuber, J., Jaczewska, J., Lekki, J., and Stachura, Z.

(2012). Cancer cell recognition: mechanical phenotype. *Micron,* **43**, pp. 1259–1266.

8. Ketene, A. N., Schmelz, E. V., Roberts, P. C., and Agah, M. (2012). The effects of cancer progression on the viscoelasticity of ovarian cell cytoskeleton structures. *Nanomed. Nanotechnol. Biol. Med.,* **8**, pp. 93–102.
9. Baker, E. L., Bonnecaze, R. T., and Zaman, M. H. (2009). Extracellular matrix stiffness and architecture govern intracellular rheology in cancer. *Biophys. J.,* **97**, pp. 1013–1021.
10. Tang, X., Kuhlenschmidt, T. B., Zhou, J., Bell, P. Wang, F., Kuhlenschmidt, M. S., and Saif, T. A. (2010). Mechanical force affects expression of an in vitro metastasis-like phenotype in HCT-8 cells. *Biophys. J.,* **99**, pp. 2460–2469.
11. Lekka, M., Lekka, M., Gil, D., Pogoda, K., Dulińska-Litewka, J., Jach, R., Gostek, J., Klymenko, O., Prauzner-Bechcicki, Sz., Stachura, Z., Wiltowska-Zuber, J., Okoń, K., and Laidler, P. (2012) Cancer cell detection in tissue sections using AFM. *Arch. Biochem. Biophys.,* **518**, pp. 151–156.
12. Plodinec, M., Loparic, M., Monnier, C. A., Obermann, E. C., Zanetti-Dallenbach, R., Oertle, P., Hyotyla, J. T., Aebi, U., Bentires-Alj, M., Lim, R. Y., and Schoenenberger, C. A. (2012). The nanomechanical signature of breast cancer. *Nat. Nanotechnol.,* **7**, pp. 757–765.

Index